M000074486

Adobe Certified

David Mayer

2020-06-11

<u>How To Pass Adobe Exams</u>

Adobe Certified

Complete Step-By-Step Guide To Quickly Pass All Adobe Exams And Improve Your Job Position

How to Get a 50% Discount

If you have come this far and are curious about how to get a unique offer, then you surely understand it is essential for you to take the right Adobe Certification now. There is no more time to waste; it's time to be certified in the right way according to your skills.

By purchasing this book you will then be entitled to an incredible 50% discount code on all products available at www.Certification-questions.com

It is a unique offer, and you will finally be able to understand which is the Certification required for your career, understand how to take the Exam and what are the prerequisites. Also, you will be able to use the only Simulated Exam that always has the latest questions available with 100% guaranteed success and a money back guarantee policy.

Don't wait, send us a copy to info@Certification-questions.com:

- of your book
- of your purchase receipt

And, we will automatically send you a discount code equal to 50% for the best Exam Simulator available on the market.

Thanks - Certification-questions.com

Introduction

Nowadays, finding good work is more difficult than ever; competition is high and skills become an essential factor to check when hiring somebody. For this reason, assuring companies that they have the right people with the right certified skills and knowledge, through Certifications, qualifications and degrees, is the best way of opening the doors to the world of work.

Adobe Certifications are not to be underestimated.

Adobe Certification, for those who are not aware of it, is an official document that enables a professional to show their certified skills and knowledge.

An Adobe Certification can be of different types, depending on the level of preparation and the professional category to which they refer. There are three levels of Adobe Certification:

- Adobe Certified Associate (ACA)
- Adobe Certified Expert (ACE)
- Adobe Certified Instructor (ACI)

Adobe Certified Associate is designed to test basic knowledge with Adobe software tools such as Photoshop, Dreamweaver, Flash, Premiere Pro, Illustrator and InDesign. The Exam is both theoretical and practical, and consists of multiple choice questions.

The Adobe Certified Expert is designed for professionals; it is an official document recognized all over the world and aimed at expert users who, upon passing the Exam, will be able to propose themselves as an experienced and qualified specialist of one or more Adobe programs. The Exam is theoretical and is carried out on an English language platform. The Adobe ACE program is much more challenging than the ACA program and requires a lot of effort and proper preparation. Therefore, it is good to study and practice as much as possible when trying for ACE Certification.

The Adobe Certified Instructor does not provide actual Certifications but instead offers qualifications, which are obtained from time to time by passing the ACE Exam together with the CompTIA CTT + (Certified Technical Trainer) qualification Exam.

This guide provides all the information you need to easily pass Adobe Exams. Whether you are a beginner or an expert, our team will provide you with everything you need to start studying in detail.

In this guide we will offer you step by step information on the following topics:

- Adobe Certification Exam registration procedure
- Adobe Certification Exam topics
- Benefits of obtaining an Adobe Certification
- Requirements for Adobe Certification Exams
- The Certification path describes the knowledge of the technologies and related skills necessary to pass the Exams
- Duration of the Exam
- Exam format
- Certified professional salary in different countries
- Price of Adobe Certification Exams
- Practical tests
- How to obtain the Certification that best suits your professional growth

- 50% discount on the purchase of the Web And Mobile Simulator available at www.Certification-questions.com
- 100% success guarantee

The guide contains everything you need to prepare for the official Adobe Exams, and to master the skills needed to obtain the Certification that best suits your career path.

In addition to that, we will also provide you with everything you need for making the right decision when choosing an Adobe Certification: The material in this guide helps professionals to choose the right Certification and to improve their knowledge to pass the official Certification Exam quickly and easily. Our team already provides selected and targeted questions \ answer content to consolidate professional preparation through practical tests and training material.

Our guide provides 100% Exam information, ensuring your preparation is solid and detailed. Thanks to this guide, you will obtain all the necessary information for choosing and passing the Certification that best suits your needs within the huge Adobe Certification Program.

Copyright

Dedication

Do you know how?

You choose a book then go to the Dedication Page and find that once again the author has dedicated a book to someone else and not to you.

Not this time.

This book is dedicated to the readers, as, without you, there would be no need for this book to have been written.

Hopefully the effort put into producing this resource guide will result in value and success when you sit your certification exam.

This one's for you. Thanks - Certification-questions.com

Acknowledgments

The world is better, thanks to people who want to develop and guide others. What makes it even better are the people who share the gift of their time to guide future leaders. Thank you to all who strive to grow and help others to grow.

Without the experience and support of my colleagues and team at Certification-questions.com, this book would not exist. You have given me the opportunity to lead a great group of people to become a leader of great leaders. It is a blessed place. Thanks to the Certification-questions.com team.

I want to say thank you to everyone who ever said anything positive to me or taught me something. It was your kind words and actions over the years that drove me to help others in my turn. THANK YOU.

PREFACE

Are you looking for valid Practice Tests for Adobe Certification?

This book will guide on how you can pass the Adobe Certification Exam using Practice Tests.

We will cover a large set of information for Adobe Certification topics, so you will systemically discover how to pass the Certification exam.

This book will also explore many of your questions, such as:

- Adobe Exam topics
- What are the essential criteria for passing the Official Adobe Exam?
- How much Adobe Exam Cost?
- What is the format of the Adobe Exam?
- The advantage of Adobe Exam Certification
- What are the difficulties of Adobe Exam Certification?

We appreciate you taking the time to read this book, and we are really excited to assist you on your career growth journey.

How to Use This Book

There are four main components to the present Adobe Study Guide.

First, the Introduction, in which you will get to know about the importance of Adobe Certification and Practice Tests.

Secondly, the Table of Contents proves quite helpful for maneuvering through the ebook.

Thirdly, there is the Content, in which you will get to know about the different methods of studying for the Adobe Certification Exam that will help you to pass the Certification Exam on your first attempt.

Fourthly, the Summary, in which you will read the brief statement or account of the main points of the Adobe Certification Exam.

Index

Adobe Exams

Adobe Certified Professionals form a unique community with Adobe as its hub.Individuals can take advantage of the networking and professional growth opportunities which according to the research is a much more poignant aspect of the value of certification that was previously envisioned. Adobe also recognizes that the community is an important way to engage with its customer base.

The **Certification-questions.com team** has worked directly with industry experts to provide you with the actual questions and answers from **the latest versions of the Adobe exam**. Practice questions are proven to be the most effectively way of preparing for certification exams.

Adobe certified professionals are certified individuals who specialize in Adobe information technology programs and applications. Experts in the field of Adobe programs, they focus their technical support skills in various areas, ranging from operating systems, cloud solutions to Web development.

With a certificate, your value increases when you apply for jobs. According to Adobe your chances of getting **hired increases 5 times**. According to Adobe, **86% of hiring managers indicate that they prefer job applicants having an IT certificate**. And Adobe certification is a preference over some unknown computer training institutes' certificates. Eight out of ten Hiring Managers wish to verify the certificates provided by job applicants. Further, according to Adobe, 64% of IT managers prefer Adobe certificates to other certificates. Certification, training, and experience are the three main areas that provide better recognition to a person when it comes to promotions and incentives.

We offers an online service that allows students to study through tests questions. The Simulator is built to reflect the final exam structure: It is an excellent study material as it offers the ability to run an online actual exam. Every question is also associated with the solution and each solution is explained in detail.

Chapter 1: 9A0-303 - Adobe Photoshop CS6 ACE

Exam Guide

Adobe Photoshop CS6 ACE 9A0-303 Exam:

Adobe Photoshop CS6 ACE 9A0-303 Exam is related to Adobe Photoshop CS6 ACE and Credits toward Adobe Photoshop Certification. This exam validates the ability to understand process version and workflow options analyzed ingested Media and determine the format needed based on intended needs apply basic tonal and color corrections to images. It also deals with the ability to organize individual Photoshop files using best practices for layer organization understanding the differences between raster and vector layers. Adobe Certified Expert Partners, Customers and Consultants usually hold or pursue this certification and you can expect the same job role after completion of this certification.

9A0-303 Exam topics:

Candidates must know the exam topics before they start of preparation. Because it will really help them in hitting the core. Our **9A0-303 dumps** will include the following topics:

- Managing Assets using Adobe Bridge
- Using Camera Raw
- Understanding Photoshop Fundamentals
- Understanding Selections

- Understanding Layers
- Understanding Adjustments
- Editing Images
- Working with Videos

Certification Path:

The Adobe Photoshop CS6 ACE certification path includes only one 9A0-303 certification exam.

Who should take the 9A0-303 exam:

The Adobe Photoshop CS6 ACE 9A0-303 Exam certification is an internationally-recognized validation that identifies persons who earn it as possessing skilled in Adobe Photoshop. If a candidate wants significant improvement in career growth needs enhanced knowledge, skills, and talents. The Adobe Photoshop CS6 ACE 9A0-303 Exam certification provides proof of this advanced knowledge and skill. If a candidate has knowledge of associated technologies and skills that are required to pass Adobe Photoshop CS6 ACE 9A0-303 Exam then he should take this exam.

How to study the 9A0-303 Exam:

There are two main types of resources for preparation of certification exams first there are the study guides and the books that are detailed and suitable for building knowledge from ground up then there are video tutorial and lectures that can somehow ease the pain of through study and are comparatively less boring for some candidates yet these demand time and concentration from the learner. Smart Candidates who want to build a solid foundation in all exam topics and related technologies usually combine video lectures with study guides to reap the benefits of both but there is one crucial preparation tool as often overlooked by most candidates the practice exams. Practice exams are built to make students comfortable with the real exam environment. Statistics have shown that most students fail not due to that preparation but due to exam anxiety the fear of the unknown. Certification-questions.com expert team recommends you to prepare some notes on these topics along with it don't forget to practice 9A0-303 Exam dumps which been written by our expert team, Both these will help you a lot to clear this exam with good marks.

How much 9A0-303 Exam Cost:

The price of 9A0-303 exam is $180 USD.

How to book the 9A0-303 Exam:

These are following steps for registering the 9A0-303 exam.

- Step 1: Visit to Pearson Exam Registration
- Step 2: Signup/Login to Pearson VUE account
- Step 3: Search for Adobe 9A0-303 Exam Certifications Exam
- Step 4: Select Date, time and confirm with payment method

What is the duration of the 9A0-303 Exam:

- Format: Multiple choices, multiple answers
- Length of Examination: 95 minutes
- Number of Questions: 50
- Passing Score: 60%

The benefit in Obtaining the 9A0-303 Exam Certification:

- Resumes with Adobe Certified Expert certifications get noticed and fast-tracked by hiring managers.
- Adobe Certified Expert recognition and respect from colleagues and employers.

- Adobe Certified Expert receives exclusive updates on Adobe's latest products and innovations.
- Adobe Certified Expert can display Adobe certification logos on their business cards, resumes, and websites to leverage the power of the Adobe brand.

Difficulty in writing 9A0-303 Exam:

Candidates face many problems when they start preparing for the 9A0-303 exam. If a candidate wants to prepare his for the 9A0-303 exam without any problem and get good grades in the exam. Then they have to choose the best 9A0-303 dumps for real exam questions practice. There are many websites that are offering the latest 9A0-303 exam questions and answers but these questions are not verified by Adobe certified experts and that's why many are failed in their just first attempt. Certification-questions is the best platform which provides the candidate with the necessary 9A0-303 questions that will help him to pass the 9A0-303 exam on the first time. Candidate will not have to take the 9A0-303 exam twice because with the help of **Adobe 9A0-303 exam dumps** Candidate will have every valuable material required to pass the Adobe 9A0-303 exam. We are providing the latest and actual questions and that is the reason why this is the one that he needs to use and there are no chances to fail when a candidate will have valid braindumps from Certification-questions. We have the guarantee that the questions that we have will be the ones that will pass candidate in the 9A0-303 exam in the very first attempt.

For more info visit::

9A0-303 Exam Reference

Sample Practice Test for 9A0-303

Question: 1 *One Answer Is Right*

Which of the following file formats is a standard Windows image format on DOS and Windows-compatible computers?

Answers:

A) BMP

B) Photoshop EPS

C) Photoshop Raw

D) Digital Negative (DNG)

Solution: A

Explanation:

Explanation: The BMP format supports RGB, Indexed Color, Grayscale, and Bitmap color modes. BMP is a standard Windows image format on DOS computers and Windows-compatible computers. Answer option B is incorrect. The Encapsulated PostScript (EPS) language file format can contain both vector and bitmap graphics. It is supported by virtually all graphics, illustration, and page-layout programs. To transfer PostScript artwork between applications, you can use the EPS format. Answer option C is incorrect. The Photoshop Raw format supports CMYK, RGB, and grayscale images with alpha channels. It supports multichannel and Lab images without alpha channels. Answer option D is incorrect. The Digital Negative (DNG) file format is used to contain the raw image data from a digital camera and metadata that defines what the data means. DNG, Adobe's publicly available, archival format for camera raw files is designed to provide compatibility and decrease the current proliferation of camera raw file formats. Reference: http://help.adobe.com/en_US/photoshop/cs/using/WSfd1234e 1c4b69f30ea53e41001031ab64- 7758a.html

Question: 2 *One Answer Is Right*

Which of the following methods are correct to open the Layer Style dialog box to add a layer style? Each correct answer represents a complete solution. Choose two.

Answers:

A) Press Alt and click the Create a new layer icon in the Layer panel.

B) In the Layers panel, select a style from the Blending mode list.

C) In the Layers panel, click the Add a layer style icon and select the layer style you want to apply.

D) Choose Layer > Layer Style, and from the submenu that appears, choose the layer style you want to apply.

Solution: C, D

Explanation:

Explanation: To open the Layer Style dialog box, you can do any of the following: - In the Layers panel, click the Add a layer style icon and select the layer style you want to apply. - Choose Layer > Layer Style. From the submenu that appears, choose the layer style you want to apply. - In the Layers panel, double-click a pixel-based layer's thumbnail, or double-click to the right of the layer name for non-pixel-based or type layers. Answer options A and B are incorrect. These methods are invalid.

Question: 3 *One Answer Is Right*

Which of the following options from the Batch dialog box prevents the display of the Camera Raw dialog box as each camera raw image is processed?

Answers:

A) Override Action "Save As" Commands

B) Override Action "Open" Commands

C) Suppress File Open Options Dialogs

D) Include All Subfolders

Solution: C

Explanation:

Explanation: To prevent the display of the Camera Raw dialog box as each camera raw image is processed, select Suppress File Open Options Dialogs while using the Batch command. Answer option B is incorrect. When you use the Batch command, select Override Action "Open" Commands. Any Open commands in the action will then operate on the batched files rather than the files specified by name in the action. Answer option A is incorrect. When you use the Batch command, select Override Action "Save As" Commands if you want to use the Save As instructions from the Batch command instead of the Save As instructions in the action. Answer option D is incorrect. Include All Subfolders is not used to save the files. Reference: http://helpx.adobe.com/pdf/bridge_reference.pdf

Question: 4 *One Answer Is Right*

While working with the Merge To HDR, which of the following tone-mapping methods adjusts the intensity of subtle colors?

Answers:

A) Color

B) Local Adaptation

C) Toning Curve

D) Edge Glow

Solution: A

Explanation:

Explanation: Color is used to adjust the intensity of subtle colors, while minimizing clipping of highly saturated colors. Saturation adjusts the intensity of all colors from -100 (monochrome) to +100 (double saturation). Answer option B is incorrect. Local Adaptation is used to adjust HDR tonality by adjusting local brightness regions throughout the image. Answer option C is incorrect. Toning Curve displays an adjustable curve over a histogram showing luminance values in the original, 32-bit HDR image. Answer option D is incorrect. The Edge Glow radius specifies the size of the local brightness regions in the image. Strength specifies how far apart two pixels' tonal values must be before they are no longer part of the same brightness region. Reference: http://help.adobe.com/en_US/photoshop/cs/using/WSfd1234e 1c4b69f30ea53e41001031ab64-78eea.html#WSfd1234e1c4b69f30ea53e41001031ab64-78e5a

Question: 5 *One Answer Is Right*

Which of the following options in the Refine Edge dialog box is used to determine the size of the selection border in which edge refinement occurs?

Answers:

A) Smart Radius

B) Radius

C) View Mode

D) Refine Radius tools

Solution: B

Explanation:

Explanation: Radius is used to determine the size of the selection border in which edge refinement occurs. Answer option C is incorrect. View Mode is used to change the mode that how the selection is displayed. Answer option D is incorrect. Refine Radius is used to precisely adjust the border area in which the edge refinement occurs. Answer option A is incorrect. Smart Radius is used to automatically adjust the radius for soft and hard edges found in the border region. Reference: http://help.adobe.com/en_US/photoshop/cs/using/WSfd1234e 1c4b69f30ea53e41001031ab64- 76f0a.html#WS9C5407FF- 2787-400b-9930-FF44266E9168a

Question: 6 *Multiple Answers Are Right*

HOTSPOT Choose the correct file format in the Format menu that supports documents up to 300,000 pixels in any dimension and all Photoshop features.

Hot Area:

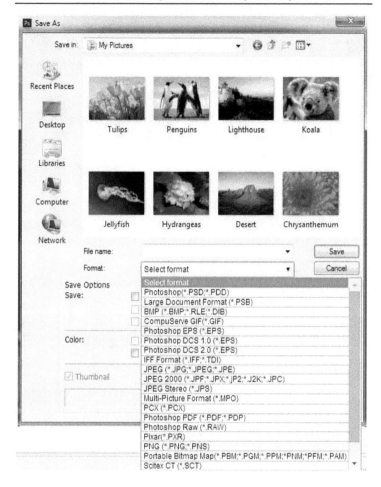

Answers:

A)

Solution: A

Explanation:

Explanation: The Large Document Format (PSB) supports documents up to 300,000 pixels in any dimension. It supports all Photoshop features, such as layers, effects, and filters.

Question: 7 *Multiple Answers Are Right*

HOTSPOT You work as a graphic designer for WEBDESIGN Inc. You are creating a vector shape in Photoshop CS6. You want to load a set of shapes from a previously saved file. Mark the correct option in the list of the shape tool. Hot Area:

Answers:

A)

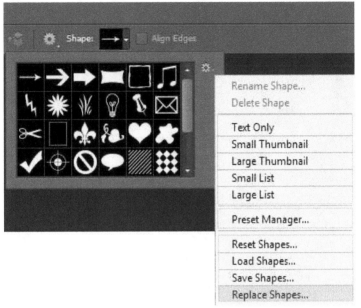

Solution: A

Explanation:

Explanation:

Question: 8 *Multiple Answers Are Right*

HOTSPOT You are working as a graphic designer for WEBDESIGN Inc. You are interpreting video footage in Photoshop CS6. You want to specify the matte color with which the channels are already multiplied. Choose the correct option in the Alpha Channel option to accomplish this task. Hot Area:

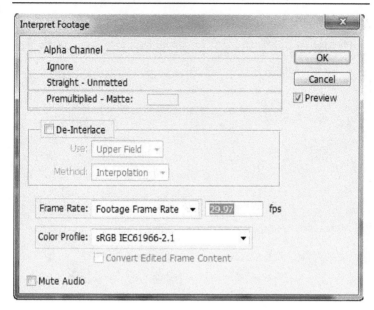

Answers:

A)

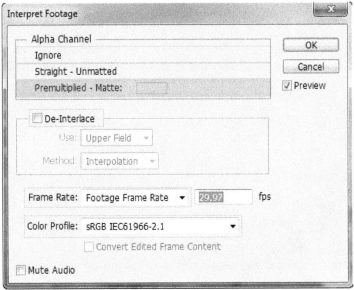

Solution: A

Explanation:

Explanation:

Question: 9 *Multiple Answers Are Right*

HOTSPOT Identify the different components of the Photoshop workspace. Hot Area:

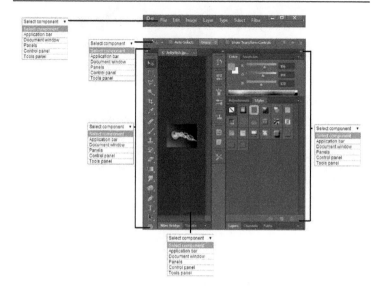

Answers:

A)

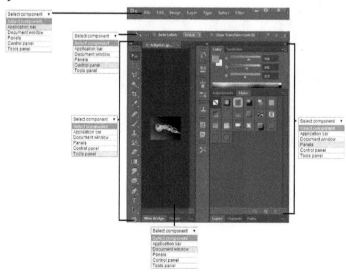

Solution: A

Explanation:

Explanation: The following table describes the components of the Photoshop workspace:

Components	Descriptions
Application bar	It contains a workspace switcher, menus (Windows only), and other application controls.
Tools panel	It contains tools for creating and editing images, artwork, page elements, and so on.
Control panel	It displays options for the currently selected tool.
Panels	It helps you monitor and modify your work.
Document window	It displays the file you are working on.

Question: 10 *Multiple Answers Are Right*

HOTSPOT You work as a graphic designer for WEBDESIGN Inc. You are using the Burn tool in Photoshop CS6. You want to change the light areas in an image. Choose the correct option in the Range menu to accomplish the task. Hot Area:

Answers:

A)

Solution: A

Explanation:

Explanation: You should choose the Highlights option from the Range menu to change the light areas in the image. The following table describes the Range menu options:

Options	Descriptions
Midtones	It is used to change the middle range of grays in an image.
Shadows	It is used to change the dark areas in an image.
Highlights	It is used to change the light areas in an image.

Chapter 2: 9A0-381 - Analytics Business Practitioner

Exam Guide

Adobe Analytics Business Practitioner 9A0-381 Exam:

Adobe Analytics Business Practitioner 9A0-381 Exam is related to Adobe Certified Expert Certification. This 9A0-381 Exam validates the ability to identify tags understand server calls and report suites view ranked versus trended reports breakdown reports and understand segmenting reports, understand the difference between dimensions and metrics. It also deals with the ability to create bookmarks and dashboards extract data add comments to reports add alerts add targets and calendar invites Candidate should also be able to use report. Senior Graphic Designers, Art Directors and Video Editors usually hold or pursue this certification and you can expect the same job roles after completion of this certification.

9A0-381 Exam topics:

Candidates must know the exam topics before they start of preparation. Because it will really help them in hitting the core. Our **9A0-381 dumps** will include the following topics:

- Conducting a business analysis 33.4%
- Reporting and Dashboarding 23.3%
- Segmenting 20%
- Administering and troubleshooting 23.3%

Certification Path:

The Adobe Analytics Business Practitioner certification path includes only one 9A0-381 certification exam.

Who should take the 9A0-381 exam:

The Adobe Analytics Business Practitioner 9A0-381 Exam certification is an internationally-recognized validation that identifies persons who earn it as possessing skilled as Adobe Certified Expert. If a candidate wants significant improvement in career growth needs enhanced knowledge, skills, and talents. The Adobe Analytics Business Practitioner 9A0-381 Exam certification provides proof of this advanced knowledge and skill. If a candidate has knowledge of associated technologies and skills that are required to pass Adobe Analytics Business Practitioner 9A0-381 Exam then he should take this exam.

How to study the 9A0-381 Exam:

There are two main types of resources for preparation of certification exams first there are the study guides and the books that are detailed and suitable for building knowledge from ground up then there are video tutorial and lectures that can somehow ease the pain of through study and are comparatively less boring for some candidates yet these demand time and concentration from the learner. Smart Candidates who want to build a solid foundation in all exam topics and related technologies usually combine video lectures with study guides to reap the benefits of both but there is one crucial preparation tool as often overlooked by most candidates the practice exams. Practice exams are built to make students comfortable with the real exam environment. Statistics have shown that most students fail not due to that preparation but due to exam anxiety the fear of the unknown. Certification-questions.com expert team recommends you to prepare some notes on these topics along with it don't forget to practice 9A0-381 Exam dumps which been written by our expert team, Both these will help you a lot to clear this exam with good marks.

How much 9A0-381 Exam Cost:

The price of 9A0-381 exam is $180 USD.

How to book the 9A0-381 Exam:

These are following steps for registering the 9A0-381 exam. Step 1: Visit to Pearson Exam Registration

- Step 2: Signup/Login to Pearson VUE account
- Step 3: Search for Adobe 9A0-381 Exam Certifications Exam
- Step 4: Select Date, time and confirm with payment method

What is the duration of the 9A0-381 Exam:

- Format: Multiple choices, multiple answers
- Length of Examination: 105 minutes
- Number of Questions: 60
- Passing Score: 63%

The benefit in Obtaining the 9A0-381 Exam Certification:

- Resumes with Adobe Certified Expert certifications get noticed and fast-tracked by hiring managers.
- Adobe Certified Expert recognition and respect from colleagues and employers.
- Adobe Certified Expert receives exclusive updates on Adobe's latest products and innovations.
- Adobe Certified Expert can display Adobe certification logos on their business cards, resumes, and websites to leverage the power of the Adobe brand.

Difficulty in writing 9A0-381 Exam:

This exam is very difficult especially for those who have not on the job experience as an Adobe Certified Expert. Candidates can not pass this exam with only taking courses because courses do not provide the knowledge and skills that are necessary to pass this exam. Certification-questions.com is the best platform for those who want to pass 9A0-381 with good grades in no time. Certification-questions.com provides the latest 9A0-381 exam dumps that will immensely help candidates to get good grades in their final 9A0-381 exam. Certification-questions.com is one of the best study sources to provide the most updated **Adobe 9A0-381 Exam Dumps** with our Actual 9A0-381 Exam Questions PDF. Candidate can rest guaranteed that they will pass their Adobe 9A0-381 Exam on the first attempt. We will also save candidates valuable time. Certification-questions Dumps help to pass the exam easily. Candidates can get all real questions from Certification-questions. One of the best parts is we also provide most updated Adobe Certified Expert Exam study materials and we also want a candidate to be able to access study materials easily whenever they want. So, We provide all our Adobe 9A0-381 exam questions in a very common PDF format that is accessible from all devices.

For more info visit::

9A0-381 Exam Reference

Sample Practice Test for 9A0-381

Question: 1 *One Answer Is Right*

You want to access Content Consumption (Page Views/Visits) per Pages, Site Sections and Site Sub Sections. You have a separate custom traffic variable reserved for each of the reports. What is the best way to ensure that the Content Consumption calculated metric is available for all three reports?

Answers:

A) You only need to build it once and it will become available for all three reports

B) Build that calculated metric three times, once in each report.

C) That metric is only available for Pages so you will be unable to apply it to Site Sections or Sub Sections

D) Choose the "copy" feature in the calculated metric builder to propagate the metric from one report to the other

Solution: A

Question: 2 *One Answer Is Right*

You are using a filter to eliminate a select group of line items. What does the metric total number represent?

Answers:

A) The total including the line items that are not filtered.

B) The site total for the selected metric

C) The percentage of the filtered line items compared against the Report Suite total

D) The total of the line items filtered

Solution: C

Explanation:

Explanation: Reference:
https://marketing.adobe.com/developer/documentation/sitecatalyst-reporting/r-metrics-1

Question: 3 *One Answer Is Right*

Which statement summarizes the reporting possibilities for subrelations?

Answers:

A) You can drill down multiple levels deep and choose up to 10 metrics for display

B) You can drill down only one level deep and choose up to 10 metrics for display

C) You can drill down only one level deep and only get to choose one metric

D) You can drill down multiple levels deep but you only get to use the Instances metric

Solution: D

Question: 4 *One Answer Is Right*

What are the maximum number of items that can be displayed in a Trended graph?

Answers:

A) 1

B) 5

C) 10

D) 30

Solution: D

Explanation:

Explanation: Reference:
https://www.linkedin.com/pulse/adobe-analytics-beginners-
post-4-report-types-swati-ramanujam

Question: 5 *One Answer Is Right*

Which statement about Classification reports is NOT true?

Answers:

A) Classified reports are used to provide friendly names to key values

B) Classified reports can be applied to every variable

C) Classified reports are controlled by an admin

D) Classified reports are used to group key values to higher level categories

Solution: B

Question: 6 *One Answer Is Right*

You are building a segment based only on Events in the Segment Canvas. Which two containers can you use? (Choose two.)

Answers:

A) Events

B) Visitors

C) Visits

D) Hits

Solution: A, C

Question: 7 *One Answer Is Right*

A visitor demonstrates the following behavior in terms of time spent:

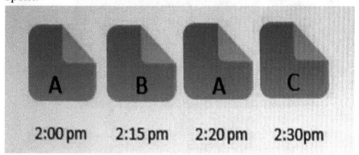

What is the average time spent on page A?

Answers:

A) 12.5 minutes

B) 25 minutes

C) 20 minutes

D) 15 minutes

Solution: D

Question: 8 *One Answer Is Right*

Click the Exhibit tab to see the exhibit.

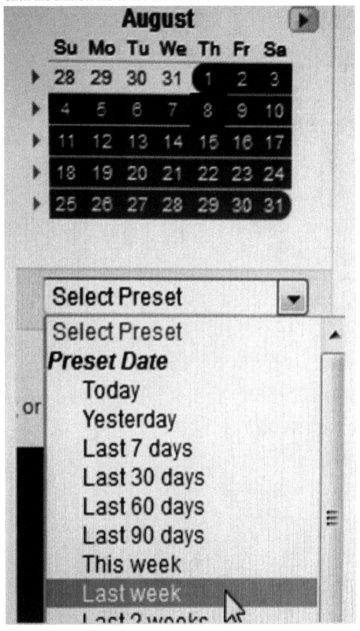

It is August 8 and you run a report for Last week. Which days will be included in the report?

Answers:

A) July 29 through August 2

B) August 2 through August 8

C) July 28 through August 3

D) August 1 through August 7

Solution: D

Question: 9 *One Answer Is Right*

Click the Exhibit button to see the exhibit.

Visit Number Report ⊕
JJ Esquire Training 20
May 2013

Visit Number	Instances		Registrations		Product Views (Custom)	
1. 3rd Visit	4,999	20.1%	286	18.5%	2,620	19.5%
2. 2nd Visit	4,522	18.2%	277	18.0%	2,488	18.5%
3. 4th Visit	4,281	17.2%	251	16.3%	2,328	17.3%
4. 5th Visit	2,954	11.9%	195	12.6%	1,685	12.5%
5. 1st Visit	2,706	10.9%	175	11.3%	1,411	10.5%
6. 6th Visit	1,693	6.8%	111	7.2%	916	6.8%
7. 7th Visit	862	3.5%	56	3.6%	484	3.6%
8. 8th Visit	371	1.5%	15	1.0%	177	1.3%
9. 9th Visit	195	0.8%	12	0.8%	105	0.8%
10. 32nd Visit	136	0.5%	6	0.4%	72	0.5%
Total	24,894		1,542		13,437	

Which report interpretation is inconclusive?

Answers:

A) Most visitors purchase products on their 3rd visit to the site.

B) 11.3 % of total registrations are completed on visitors' 1st visit to the site.

C) Most registrations are completed during visitors' 3rd visit to the site.

D) Most visitors wait until their 3rd visit to the site to convert.

Solution: A

Question: 10 *One Answer Is Right*

How many unique line items are accessible in a ranked report every calendar month?

Answers:

A) 50,000

B) 100,000

C) 250,000

D) 500,000

Solution: A

Explanation:

Explanation: Reference:
https://marketing.adobe.com/resources/help/en_US/sc/user/
analytics_reports_user.pdf (page 20)

Chapter 3: 9A0-384 - Adobe Experience Manager 6 Developer

Exam Guide

Adobe Experience Manager 6 Developer 9A0-384 Exam:

Adobe Experience Manager 6 Developer 9A0-384 Exam is a certification exam for Adobe Experience Manager 6 Developers. This exam is related to the Adobe Certified Expert certification. 9A0-348 Exam validates the ability of candidates to install Adobe Experience Manager (AEM) on supported Operating Systems with different run modes, configure application agents a Web Server and the Dispatcher, configure a source control system that can be used to manage files in Adobe Experience Manager (AEM) and deploy AEM projects by using Maven. Candidates also have the ability to create custom components, dialogues, templates, page components, and custom log files by using the web console. This exam also test candidates have knowledge of how to configure and manage AEM log level for a specific Adobe Experience Manager environment.

9A0-348 Exam topics:

Candidates must know the exam topics before they start of preparation. Because it will really help them in hitting the core. Our **9A0-348 dumps** will include the following topics:

- Installing and configuring an AEM developer environment: 10%
- Building and deploying AEM projects: 12%
- Building AEM components: 29%
- Building OSGi services: 31%
- Troubleshooting AEM projects: 18%

  Certification Path:

The Adobe Experience Manager 6 Developer 9A0-384 Exam certification path includes only one 9A0-384 certification exam.

Who should take the 9A0-384 exam:

The Adobe Certified Expert - AEM Developer certification is an internationally-recognized validation that identifies persons who earn it as possessing skilled in Adobe Experience Manager. If a candidate wants significant improvement in career growth needs enhanced knowledge, skills, and talents. The Adobe Experience Manager 6 Developer 9A0-384 Exam certification provides proof of this advanced knowledge and skill. If a person has the following necessary knowledge, skills required of an Adobe Experience Manager 6 Developer 9A0-384 Exam then he should take this exam.

- Convert AEM User Stories into an implementation approach
- Design OSGi services and servlets using an IDE with Maven
- Configure OSGi services by using the Web console
- Control OSGi configurations within the CRX repository
- Design AEM templates and page components with custom dialogs
- Design AEM custom components with dialogs
- Settings up clients libs
- Settings up a developer author instance

- Settings up a developer publish instance
- Create and manage CRX packages
- Import and export code from a JCR to a file system by using VLT
- Configure AEM workflows
- Troubleshoot and resolve issues with the local environment

How to study the 9A0-384 Exam:

Certification-questions.com expert team recommends you to prepare some notes on these topics along with it don't forget to practice **Adobe Certified Expert - Adobe Experience Manager 6 Developer 9A0-384 Dumps** which been written by our expert team, Both these will help you a lot to clear this exam with good marks.

- install AEM on supported operating systems with the different run modes.
- start AEM in debug mode for remote debugging.
- setup and configure replication agents.
  -setup and configure a Web server.
  -setup and manage OSGi configurations.
  -manage users and groups
  -manage the Access Control Level (ACL) permissions.
  -setup a source control system to manage files in AEM.
  -build and deploy AEM projects by using Maven.
  -develop custom components and dialogs
  -develop templates and page components.
  -develop client libraries.
  -develop custom OSGi services.
  -develop and manage custom OSGi configurations.
  -setup and manage OSGi services and bundles by using the Felix web console.
  -manage Maven dependencies.

-develop custom log files by using the Web console.

-setup and manage AEM log levels for specific AEM environments.

-fixing caching issues related to the Dispatcher and browsers.

-fixing AEM configurations.

How much 9A0-384 Exam Cost:

The price of 9A0-384 exam is $180 USD.

How to book the 9A0-384 Exam:

These are following steps for registering the 9A0-384 exam.

Step 1: Visit to Pearson Exam Registration

Step 2: Signup/Login to Pearson VUE account

Step 3: Search for Adobe 9A0-384 Exam Certifications Exam

Step 4: Select Date, time and confirm with payment method

What is the duration of the 9A0-384 Exam:

Format: Multiple choices, multiple answers

Length of Examination: 90 minutes

Number of Questions: 51

Passing Score: 63%

The benefit in Obtaining the 9A0-384 Exam Certification:

-Adobe Experience Manager 6 Developer 9A0-384 Exam certification is the absolute best way to communicate your proficiency in the Adobe Experience Manager.

- Resumes with Adobe Experience Manager 6 Developer certifications get noticed and fast-tracked by hiring managers.

- Adobe Certified Expert - Adobe Experience Manager 6 Developer recognition and respect from colleagues and employers.

- Adobe Certified Expert - Adobe Experience Manager 6 Developer receives exclusive updates on Adobe's latest products and innovations.

- Adobe Certified Expert - Adobe Experience Manager 6 Developer can display Adobe certification logos on their business cards, resumes, and websites to leverage the power of the Adobe brand.

Difficulty in writing 9A0-384 Exam:

This exam is very difficult especially for those who have not on the job experience as an Adobe Experience Manager Developer. Candidates can not pass this exam with only taking courses because courses do not provide the knowledge and skills that are necessary to pass this exam.

Candidates can only get good grades in the 9A0-384 exam by dedication, hard work, and most accurate preparation material. There are many online platforms which are providing 9A0-384 exam preparation material but they are not verified by experts. So, candidates have to choose a platform which gives them the best and authentic **9A0-384 dumps** which can make a good impact on your final result. There are many people rely on such kind of platforms but in the end, they are mostly getting poor grades. Candidates don't have to worry about this as Certification-questions.com is only one of the best platform that provides the best 9A0-384 exam preparation material. Our **9A0-384 dumps** consist of all the topics and the questions that will be asked in the real exam and the best part is that we provide **9A0-384 dumps** in PDF format that you can easily read it offline on smartphones and other electronic accessories such as laptops, desktops, and tablets. Certification-questions.com also contain braindumps which will be really helpful in making notes.

For more info visit::

9A0-384 Exam Reference

9A0-384 Exam ReferenceAdobe Experience Manger Help

Adobe Experience Manger Authoring documentation

Sample Practice Test for 9A0-384

Question: 1 *One Answer Is Right*

Which property should be used to find the repository where the configuration changes made in the Web Console are saved?

Answers:

A) Persistent Identity (PID)

B) Reference Repository

C) Component.id

D) Component.name

Solution: A

Explanation:

Explanation: Reference: http://docs.adobe.com/docs/en/cq/5-6-1/deploying/configuring_osgi.html

Question: 2 *One Answer Is Right*

How do you disable the "Target" context menu item on components in AEM 6.0?

Answers:

A) Set the property "cq:disable Targeting" to true on the dialog node.

B) Set the property "cq:disable Targeting" to true on thecomponent node.

C) Set the property "cq:disable Targeting" to true on the cq:editConfig node.

D) Set the property "cq:disable Targeting" to true on thecq:editConfig/cq:listeners node.

Solution: C

Explanation:

Explanation: Reference: http://docs.adobe.com/docs/en/cq/5-6-1/developing/components.html

Question: 3 *One Answer Is Right*

Which statement about the usage of declarative services while creating an OSGi component is true?

Answers:

A) ©Property annotation is used to reference to other services from the component by component runtime

B) ©Reference annotation is optional and used to define properties available to the component

C) ©Component annotation is the only required annotation and missing which will NOT declare the class as component

D) ©Service annotation is required and describes which service Interface Is served by the component

Solution: C

Explanation:

Explanation: Reference:
http://felix.apache.org/documentation/subprojects/apache-felix-maven-scr-plugin/scr-annotations.html

Question: 4 *One Answer Is Right*

Which three statements about replicate permissions on a resource are true? (Choose three.)

Answers:

A) The replication rights are evaluated bottom-up in the node tree.

B) The allow replication rights has higher precedence than deny replication rights.

C) To grant replication rights to a user on any resource, user has to have read permissions for/etc/replication, / bin.

D) The allow replication rights has lower precedence than deny replication rights.

Solution: A, B, C

Explanation:

Explanation: Reference: https://helpx.adobe.com/experience-manager/6-3/sites/administering/using/security.html

Question: 5 *One Answer Is Right*

What are three causes if Dispatcher stopped updating cache files in the cache directory on the Webserver? (Choose three.)

Answers:

A) The request to the page in question contain query string parameters.

B) The request to the page in question has authorization headers and dispatcher.any does not contain \allowAuthorized.

C) Dispatcher Flush agent is disabled on publish instance.

D) The request URI of the page in question should always start with /content.

Solution: A, B, C

Explanation:

Explanation: Reference: https://helpx.adobe.com/experience-manager/dispatcher/using/dispatcher-configuration.html

Question: 6 *One Answer Is Right*

In a typical author, publish and dispatcher setup, where is the dispatcher flush configured?

Answers:

A) In the author instance.

B) In the publish instance.

C) In the dispatcher module.

D) At the webserver level.

Solution: A

Explanation:

Explanation: Reference: https://docs.adobe.com/docs/en/dispatcher.html

Question: 7 *One Answer Is Right*

Which jar name can NOT be used to install an AEM publish instance?

Answers:

A) aem-publish-p4503, jar

B) cq5-publish-4505, jar

C) cq5-publish-4503, jar

D) cq5-publish_4503, jar

Solution: C

Explanation:

Explanation: Reference: http://docs.adobe.com/docs/en/cq/5-6-1/getting_started/download_and_startworking.html

Question: 8 *One Answer Is Right*

Which property is deprecated while resolving a template (T) that can be used as a template for page (P)?

Answers:

A) cq:allowedTemplates property of the jcr:contentsubnode of P or an ancestor of P

B) allowedPaths property of T

C) allowedParents property of T

D) allowedChildren property of the template of P

Solution: B

Question: 9 *One Answer Is Right*

Assume there are multiple publish instances (publ,pub2 and pub3) serving requests for an online shopping site. The end user is allowed to provide reviews and comments for each product and about their shopping experience. The Dispatcher module is in place to load balance the requests to publish instances and there is only one author instance, named author, where content editors create the pages. When a user, named User A, provides comments and the request being served by publish Instance publ, in which three ways are these comments replicated to pub2 and pub3? (Choose three.)

Answers:

A) Configure the dispatcher flush for the author pointing to a webserver uri on which the Dispatcheris deployed and configured.

B) Configure replication agents for the author pointing to publ, pub2 and pub3.

C) Configure reverse replication agents for the author pointing to publ, pub2 and pub3.

D) Configure a static agent for the author.

Solution: A, C, D

Explanation:

Explanation: Reference: http://docs.adobe.com/docs/en/cq/5-4/deploying/configuring_cq.html

Question: 10 *One Answer Is Right*

Which template allowed Paths expression would allow a page to be created with the path/ content / main / page1/ page2?

Answers:

A) /content/main/?

B) /content/main/[^/]+(/,*)?

C) /content/main/[A/]+[A/]

D) /content/main/*

Solution: B

Chapter 4: 9A0-385 - Adobe Experience Manager 6.0 Architect

Exam Guide

Adobe Experience Manager 6.0 Architect 9A0-385 Exam:

Adobe Experience Manager 6.0 Architect 9A0-385 Exam is a certification exam for Adobe Experience Manager 6 Developers. This exam is related to the Adobe Certified Expert certification. 9A0-385 Exam validates the ability of candidates to review all pages that are needed on a website, client systems and 3rd-party software services in which AEM will be integrated. This exam also verify the candidates have skills to analyze business processes, workflows, and requirements and have the knowledge to document the migration path for content from AS-IS to TO-BE, best practices for performance requirements, an overall security model, the translation process and driving verification by end users.

9A0-385 Exam topics:

Candidates must know the exam topics before they start of preparation. Because it will really help them in hitting the core. Our **9A0-385 dumps** will include the following topics:

- Discovering existing environments and business processes: 10%

- Discovering client expectations: 10%
- Validating business requirements: 6%
- Designing a solution architecture: 2%
- Identifying components and templates for web pages: 10%
- Creating migration strategies: 8%
- Identifying and recommending performance requirements: 8%
- Identifying and recommending a security model: 10%
- Identifying quality assurance requirements and planning the QA process.: 8%
- Integrating with third-party systems: 8%
- Managing the content editing process: 10%
- Creating the development process: 10%

Certification Path:

The Adobe Experience Manager 6.0 Architect 9A0-385 Exam certification path includes only one 9A0-385 certification exam.

Who should take the 9A0-385 exam:

The Adobe Certified Expert - AEM Architect certification is an internationally-recognized validation that identifies persons who earn it as possessing skilled in Adobe Experience Manager. If a candidate wants significant improvement in career growth needs enhanced knowledge, skills, and talents. The Adobe Experience Manager 6.0 Architect 9A0-385 Exam certification provides proof of this advanced knowledge and skill. If a person has the following necessary knowledge, skills required of an Adobe Experience Manager 6.0 Architect 9A0-385 Exam then he should take this exam.

- Define a content model for content/assets that can accommodate future requirements

- Defines the necessary templates and components based on business requirements
- Identify when to use out-of-the-box components versus custom components
- Create specifications for custom components
- Identify when to use OSGi bundles/services and tag libraries
- Identify when a proof of concept (POC) is needed
- Explain common security models and concepts (for example LDAP, SSO)
- Determines which storage type to use based upon non-functional business requirements.
- Explain and can apply common performance optimization concepts to customer requirements (for example: caching/CDN, user-generated content)
- Leads the design of workflow models for business processes
- Defines analytics tagging architecture for customer websites
- Creates a security concept for an AEM installation (Users, groups, ACLs, Dispatcher rules, OS-based security)
- Develops migration strategies from legacy systems to AEM
- Explain localization concepts that could impact content structure, templates, and components (i18n), and 3rd-party translations
- Demonstrate knowledge of User Interface (UI) frameworks
- Set up a sufficient testing infrastructure (this includes staging environments, coordination of tests, clarification of which automated and/or manual tests are mandatory)

How to study the 9A0-385 Exam:

Certification-questions.com expert team recommends you to prepare some notes on these topics along with it don't forget to practice **Adobe Certified Expert - Adobe Experience Manager 6.0 Architect 9A0-385 Dumps** which been written by our expert team, Both these will help you a lot to clear this exam with good marks.

- Analyze client systems in which AEM will be integrated
- Analyze 3rd-party software services in which AEM will be
- Examine business processes and workflows
- Examine business requirements
- Understand short-term and long-term client needs
- The report which business requirements map to AEM out-of-the-box functionality
- Qualify customer technical requirement definitions based on project progress.
- Check all pages that are needed on a website
- Classify web pages
- Create templates for categories
- Recognize out-of-the-box components to support a template
- Determine custom components that support a template
- Draw attributes from and AS-IS content structure to a TO-BE content structure
- Report the migration path for content from AS-IS to TO-BE
- Think the execution of migration strategies
- Define the feasibility of automatic and manual migration strategies
- Get customer historical performance metrics and documenting performance acceptance criteria/performance KPIs (getting from the client).
- Report best practices for performance requirements.
- Recognize client security requirements.

- Report an overall security model.
- Program QA phases based on customer requirements.
- know key areas to focus on QA.
- know target metrics for QA phases.
- Recognize considerations associated when integrating with other Adobe solutions.
- know gaps between required features and out-of-the-box features.
- Determine the toolchain that will be used on a project. (Including Version Control, Continuous Integration environment, documentation location).

How much 9A0-385 Exam Cost:

The price of the 9A0-385 exam is $180 USD.

How to book the 9A0-385 Exam:

These are following steps for registering the 9A0-385 exam.
Step 1: Visit to Pearson Exam Registration
Step 2: Signup/Login to Pearson VUE account
Step 3: Search for Adobe 9A0-385 exam Certifications Exam
Step 4: Select Date, time and confirm with payment method

What is the duration of the 9A0-385 Exam:

Format: Multiple choices, multiple answers
Length of Examination: 95 minutes
Number of Questions: 50
Passing Score: 63%

The benefit in Obtaining the 9A0-385 Exam Certification:

- Adobe Experience Manager 6.0 Architect 9A0-385 Exam certification is the absolute best way to communicate your proficiency in the Adobe Experience Manager.

- Resumes with Adobe Experience Manager 6.0 Architect certifications get noticed and fast-tracked by hiring managers.

- Adobe Certified Expert - Adobe Experience Manager 6.0 Architect recognition and respect from colleagues and employers.

- Adobe Certified Expert - Adobe Experience Manager 6.0 Architect receives exclusive updates on Adobe's latest products and innovations.

- Adobe Certified Expert - Adobe Experience Manager 6.0 Architect can display Adobe certification logos on their business cards, resumes, and websites to leverage the power of the Adobe brand.

Difficulty in writing 9A0-385 Exam:

This exam is very difficult especially for those who have not on the job experience as an Adobe Experience Manager Developer. Candidates can not pass this exam with only taking courses because courses do not provide the knowledge and skills that are necessary to pass this exam.

Candidates can only get good grades in the 9A0-385 exam by dedication, hard work, and most accurate preparation material. There are many online platforms which are providing 9A0-385 exam preparation material but they are not verified by experts. So, candidates have to choose a platform which gives them the best and authentic **9A0-385 dumps** which can make a good impact on your final result. There are many people rely on such kind of platforms but in the end, they are mostly getting poor grades. Candidates don't have to worry about this as Certification-questions.com is only one of the best platform that provides the best 9A0-385 exam preparation material. Our **9A0-385 dumps** consist of all the topics and the questions that will be asked in the real exam and the best part is that we provide **9A0-385 dumps** in PDF format that you can easily read it offline on smartphones and other electronic accessories such as laptops, desktops, and tablets. Certification-questions.com also contain braindumps which will be really helpful in making notes.

For more info visit::

9A0-385 Exam Reference
9A0-385 Exam ReferenceAdobe Experience Manger Help
Adobe Experience Manger Authoring documentation

Sample Practice Test for 9A0-385

Question: 1 *One Answer Is Right*

A company's news media website becomes unresponsive during an off-peak time. The AEM team investigates and identifies a DoS (Denial of Service) attack. Which two security measures should the architect use to reduce the chances of another attack? (Choose two.)

Answers:

A) Change the passwords for default out-of-the box users (admin, author)

B) Uninstall all sample content and users that ship out of the box with AEM

C) Disable the default selectors .feed.xml and .infinity.json

D) Implement a Dispatcher filter to allow only known selectors

Solution: C, D

Explanation:

Explanation: Reference:
http://docs.adobe.com/docs/en/dispatcher/disp-config.html

Question: 2 *One Answer Is Right*

A customer plans to display icons next to the search results for keyword queries on their site. These icons should be based on the type of document that results from the search. Which step should the architect recommend?

Answers:

A) Copy the entire foundation search component into the project under /apps and modify the search.jsp

B) Create a matching folder structure under /apps for the search component with a custom search.jsp

C) Create a new component under the /apps project called search with a sling:resourceSuperType of the foundation search component

D) Make modifications the search.jsp for the foundation component under /libs

Solution: C

Explanation:

Explanation: Reference:
http://docs.adobe.com/docs/en/aem/6-0/develop/components.html

Question: 3 *One Answer Is Right*

A large retail company has integrated AEM into its commerce stack. Logged in users can submit product ratings through the ratings" component. To ensure initial ratings do not skew results too far positive or negative, the company plans to show the tallied ratings only after a minimum number of submissions has occurred. For which reason must the ratings component be extended to meet these requirements?

Answers:

A) Only the average rating is available via API

B) The Tally/Rating Component API does not provide the number of user responses.

C) The author must be able to provide the minimum number of votes for display.

D) A custom rating service must be created to support this functionality.

Solution: B

Explanation:

Explanation: Reference: http://docs.adobe.com/docs/en/aem/6-0/develop/social-communities/scf.html

Question: 4 *One Answer Is Right*

A large telecommunications company is leveraging the capability of AEM to integrate with SAML. The architect must configure the SAML Authentication Handler. By default the SAML 2.0 Authentication handler is disabled. Which two OSGi configuration properties must the architect set to enable the SAML 2.0 Authentication handler? (Choose two.)

Answers:

A) The public certificate of the identity provider

B) The identity provider POST URL

C) The Service Provide Entity ID

D) The Service Ranking

Solution: B, C

Explanation:

Explanation: Reference:
http://docs.adobe.com/docs/en/dispatcher/disp-config.html

Question: 5 *One Answer Is Right*

A customer is introducing blog and forum functionality. To handle the moderation of data between the publisher and author instances, the customer needs to add reverse replication agents. The architect must create documentation based on a proposed solution architecture. Which two architectural diagrams should the architect modify?

Answers:

A) Logical

B) Physical

C) Conceptual

D) Data Flow

Solution: A, B

Question: 6 *One Answer Is Right*

AEM installs default groups and users. Which two concepts should the architect know about the admin, anonymous, and author users? (Choose two.)

Answers:

A) Keep all three users as default

B) All default accounts should be deleted.

C) Modifying the anonymous account creates additional security implications.

D) Change the password for the admin account from the default.

Solution: B, C

Explanation:

Explanation: Reference:
http://docs.adobe.com/docs/en/cq/current/administering/security.html

Question: 7 *One Answer Is Right*

A news company uses AEM to manage content. The company initially plans to provide news only to subscribers within Australia. Six months later, the company expands into the wider Asia region and requires Chinese and Korean language options. The architect must create a solution for a CMS with a multi-lingual requirement. What is the most important factor for the architect to consider?

Answers:

A) Creating a new language location for the dictionary

B) Translation workflows

C) Maintaining a common look and feel

D) Site structure

Solution: B

Question: 8 *One Answer Is Right*

An architect is integrating a customer's AEM authoring system to authenticate with the company's Active Directory (AD) directory service. The architect must configure the LoginModule used by the identity provider. Which Java security framework should the architect use?

Answers:

A) OWASP

B) PAM

C) JAAS

D) Apache Shiro

Solution: C

Explanation:

Explanation: Reference:
http://docs.adobe.com/docs/en/dispatcher/disp-config.html

Question: 9 *One Answer Is Right*

An architect needs to document the flow of content from a system that provides real-time stock quotes to an AFM system for display. The documentation must show how each server communicates and which ports are used. Which architectural diagram should the architect modify?

Answers:

A) Conceptual

B) Physical

C) Logical

D) Data Flow

Solution: A

Question: 10 *One Answer Is Right*

A client installs AEM using an application server. The architect must implement HTTPS in the client's QA environment with an application server as well. Which three AEM components/configurations must be changed relative to the QA environment? (Choose three.)

Answers:

A) Add secure flag to dispatcher configuration file

B) Install the SSL version of the dispatcher

C) Configure the CQ servlet engine HTTP service to use SSL using the company certificate

D) Configure replication agents to use SSL

Solution: A, B, D

Chapter 5: 9A0-388 - Adobe Experience Manager 6 Business Practitioner

Exam Guide

Adobe Experience Manager 6 Business Practitioner 9A0-388 Exam:

Adobe Experience Manager 6 Business Practitioner 9A0-388 Exam is a certification exam for Adobe Experience Manager 6 Business Practitioner. This exam is related to the Adobe Certified Expert certification. 9A0-388 Exam is targeted toward product specialists who understand all of the product features and functions of the product. A typical candidate has 1+ years' experience working with AEM digital assets manager, 1+ years' experience working with AEM web content management capabilities, and has 1+ years' experience working with the AEM mobile capabilities.

9A0-388 Exam topics:

Candidates must know the exam topics before they start of preparation. Because it will really help them in hitting the core. Our **9A0-388 dumps** will include the following topics:

- Understanding digital marketing concepts: 27%
- Building and managing a website with AEM: 22%
- Working with web content management tools: 34%

- Working with digital asset management tools: 17%

    Certification Path:

The Adobe Experience Manager 6 Business Practitioner 9A0-388 Exam certification path includes only one 9A0-388 certification exam.

Who should take the 9A0-388 exam:

The Adobe Certified Expert - AEM Practitioner certification is an internationally-recognized validation that identifies persons who earn it as possessing skilled in Adobe Experience Manager. If a candidate wants significant improvement in career growth needs enhanced knowledge, skills, and talents. The Adobe Experience Manager 6 Business Practitioner 9A0-388 Exam certification provides proof of this advanced knowledge and skill. If a person has the following necessary knowledge, skills required of an Adobe Experience Manager 6 Business Practitioner 9A0-388 Exam then he should take this exam.

- Determine Use Cases
- Design business requirements documents that developers can use in the creation of an AEM website
- Order digital assets
- Lead developers on the integration needed to meet business objectives
- Work on tagging to meet business requirements
- Run multiple micro-sites
- Explore all of the mobile capabilities of AEM including websites, apps, and multi-site management
- Create web pages based on the components that developers have created
- Define how AEM will integrate with other solutions
- Use workflows

- Create segments
- Control Versions

How to study the 9A0-388 Exam:

Certification-questions.com expert team recommends you to prepare some notes on these topics along with it don't forget to practice **Adobe Experience Manager 6 Business Practitioner 9A0-388 Dumps** which been written by our expert team, Both these will help you a lot to clear this exam with good marks.

- learn Social Media integration
- learn personalization
- learn analytics
- learn external data that is needed for a website
- learn Social Communities (Blogs, Forums, etc.)
- learn versioning web content
- learn the versioning of digital assets
- learn reporting
- control microsites
- handle multi-lingual sites
- learn the Sidekick
- learn how to build pages
- learn templates
- have a basic Knowledge of AEM architecture
- fix and resolve issues related to missing content or pages not updating
- learn AEM Digital Assets Manager

How much 9A0-388 Exam Cost:

The price of the 9A0-388 exam is $180 USD.

How to book the 9A0-388 Exam:

These are following steps for registering the 9A0-388 exam.

Step 1: Visit to Pearson Exam Registration

Step 2: Signup/Login to Pearson VUE account

Step 3: Search for Adobe 9A0-388 exam Certifications Exam

Step 4: Select Date, time and confirm with payment method

What is the duration of the 9A0-388 Exam:

Format: Multiple choices, multiple answers

Length of Examination: 55 minutes

Number of Questions: 40

Passing Score: 63%

The benefit in Obtaining the 9A0-388 Exam Certification:

- Adobe Experience Manager 6 Business Practitioner 9A0-388 Exam certification is the absolute best way to communicate your proficiency in the Adobe Experience Manager.

- Resumes with Adobe Experience Manager 6 Business Practitioner certifications get noticed and fast-tracked by hiring managers.

- Adobe Certified Expert - Adobe Experience Manager 6 Business Practitioner recognition and respect from colleagues and employers.

- Adobe Certified Expert - Adobe Experience Manager 6 Business Practitioner receives exclusive updates on Adobe's latest products and innovations.

- Adobe Certified Expert - Adobe Experience Manager 6 Business Practitioner can display Adobe certification logos on their business cards, resumes, and websites to leverage the power of the Adobe brand.

Difficulty in writing 9A0-388 Exam:

This exam is very difficult especially for those who have not on the job experience as an Adobe Experience Manager Architect. Candidates can not pass this exam with only taking courses because courses do not provide the knowledge and skills that are necessary to pass this exam. Candidates can only get good grades in the 9A0-388 exam by dedication, hard work, and most accurate preparation material. There are many online platforms which are providing 9A0-388 exam preparation material but they are not verified by experts. So, candidates have to choose a platform which gives them the best and authentic **9A0-388 dumps** which can make a good impact on your final result. There are many people rely on such kind of platforms but in the end, they are mostly getting poor grades. Candidates don't have to worry about this as Certification-questions.com is only one of the best platform that provides the best 9A0-388 exam preparation material. Our **9A0-388 dumps** consist of all the topics and the questions that will be asked in the real exam and the best part is that we provide **9A0-388 dumps** in PDF format that you can easily read it offline on smartphones and other electronic accessories such as laptops, desktops, and tablets. Certification-questions.com also contain braindumps which will be really helpful in making notes.

For more info visit::

9A0-388 Exam Reference
9A0-388 Exam ReferenceAdobe Experience Manger Help
Adobe Experience Manger Authoring documentation

Sample Practice Test for 9A0-388

Question: 1 *One Answer Is Right*

What is the best way to ensure a component appears on every single page created with a template? (Choose the best answer.)

Answers:

A) Use multiple paragraph systems

B) Be sure the component is included in the page component

C) Train the authors to add the component to each page every time

D) Add the component to the paragraph system

Solution: D

Explanation:

Explanation: When enabled and located on your page you can then use Design mode to configure aspects of the component design by editing the attribute parameters. This actually involves adding, or removing, the components allowed in the paragraph system for the page. The paragraph system (parsys) is a compound component that contains all other paragraph components. The paragraph system allows authors to add components of different types to a page as it contains all other paragraph components. Each paragraph type is represented as a component. References: https://docs.adobe.com/docs/en/aem/6-2/author/page-authoring/default-components/ designmode.html

Question: 2 *One Answer Is Right*

Which does Adobe AEM Assets create when creating a version of an asset? (Choose the best answer.)

Answers:

A) Sub assets only

B) Metadata only

C) Renditions only

D) Renditions, metadata and sub assets

Solution: B

Explanation:

Explanation: Metadata is automatically versioned together with the corresponding asset. You can import or export asset metadata. Examples of when you might create versions include the following scenarios: You upload a file to CQ DAM. For example, if you modify a CQ5 DAM asset externally in another program and upload it through the user interface or WebDAV, CQ5 creates a new version of that asset so your original image is not overwritten. You edit the metadata. References: https://docs.adobe.com/docs/en/cq/5-6-1/dam/dam_documentation.html

Question: 3 *One Answer Is Right*

Which is the best definition for a rollout configuration? (Choose the best answer.)

Answers:

A) Indicates the pages available on a publish server

B) Indicates all of the components an AEM site will use

C) Indicates all of the templates an AEM site will use

D) Defining an action will be performed on a live copy page

Solution: D

Explanation:

Explanation: For each Live Copy, a Rollout Config determines how content is automatically updated. A Rollout Config consists of the following items: An event that triggers the update, such as a change to the content of the source page. One or more actions that occur, such as updating the content on the Live Copy. References: https://docs.adobe.com/docs/en/cq/5-6-1/administering/multi_site_manager.html

Question: 4 *One Answer Is Right*

In order to build a landing page in Adobe Experience Manager, you should: (Choose the best answer.)

Answers:

A) AEM does not support landing pages

B) Install the landing page package

C) Use the landing page template

D) Create an AEM page with a blank canvas

Solution: B

Explanation:

Explanation: In AEM, you create landing pages by performing the following steps: 1. Create a page in AEM that contains the landing pages canvas. AEM ships with a sample called "Blank Canvas Page." 2. Prepare the HTML and assets. 3. Package the resources into a ZIP file referred to here as the "Design

Package." 4. Import the design package on the page with the canvas. 5. Modify and publish the page. Note: With Experience Manager, marketers can take advantage of the web content management system to engage with customers across interconnected digital properties, including landing pages, emails, websites, mobile sites, and more. References: https://docs.adobe.com/docs/en/cq/5-6-1/wcm/campaigns/landingpages.html

Question: 5 *One Answer Is Right*

What kind of business goals can be achieved by implementing social media plugins into a page? (Choose the best answer.)

Answers:

A) It is an easy way for the user to get new followers on Twitter.

B) Sharing products on Twitter generates direct revenue for the marketer automatically.

C) By sharing a desired product on Facebook, the product will appear automatically on price comparing websites.

D) By sharing a product page of a webshop on Twitter, the followers and friends of the user will get the product easily recognized, which increases awareness.

Solution: D

Explanation:

Explanation: Using social plugins in CQ Social Communities, marketers can strengthen the connection between their digital properties and social networks by allowing customers to share content with friends. With new social plugins in CQ Social Communities, including the "Like" button and activity feeds, marketers can strengthen the connection between their digital properties and social networks by allowing customers to share content with friends. Organizations can also provide consumers socially relevant information on the company's website—for example, which friends purchased a product—while amplifying their presence back to the social network. Note: Social Plugins include: - Facebook Activity Feed - Facebook Comments - Facebook Facepile - Facebook Like Button - Facebook Live Stream - Facebook Send Button - Twitter Follow Button - Twitter Search - Twitter Share Button References: http://www.adobe.com/aboutadobe/pressroom/ pressreleases/201205/051512AdobeCQSocialCommunities.ht ml

Question: 6 *One Answer Is Right*

Mow con forums be moderated in Adobe Experience Manager? (Choose the best answer.)

Answers:

A) Forums cannot be moderated in Adobe Experience Manager

B) Through the notification system

C) Developers would have to create a custom moderation system

D) Using the community console

Solution: D

Explanation:

Explanation: Moderation of user generated content (UGC) is useful for recognizing positive contributions as well as limiting negative ones (such as spam and abusive language). UGC can be moderated from several environments: - bulk moderation console The Moderation console is accessible by administrators and community moderators in the public environment as well as by administrators in the author environment. This is possible when community content is stored in a common store. - in-context moderation Moderation in the publish environment may be performed by administrators and community moderators directly on the page where the content was posted. References: https://docs.adobe.com/docs/en/aem/6-2/administer/communities/moderate-ugc.html

Question: 7 *One Answer Is Right*

What is the best way to set up an approval system for a new digital asset? (Choose the best answer.)

Answers:

A) Use Adobe Drive

B) Use a Workflow

C) Use Adobe Bridge

D) Create a custom component

Solution: B

Explanation:

Explanation: Workflow-driven asset handling Create, maintain and control your workflows using an easy drag-and-drop interface, allowing you to attach automatic processing and media conversion tasks to any assets. Projects, workflow and task management With projects and task management, easily plan, review, approve and manage the production of marketing assets — even when you're juggling dozens of projects and working with multiple teams. References: http://www.adobe.com/ca/marketing-cloud/enterprise-content-management/digital-asset- workflow.html

Question: 8 *One Answer Is Right*

What is the best way to add personalization keywords to the client context cloud? (Choose the best answer.)

Answers:

A) Use a teaser component

B) Use a segmentation rule

C) Use a workflow step

D) Use the generic store properties component

Solution: D

Explanation:

Explanation: Customized client context is used to re use the personalization according to our requirements. Usually people extend the 'Generic store property' to customize the personalization. References: http://aem-cq-tutorials.blogspot.se/

Question: 9 *One Answer Is Right*

What is the best way to reduce the number of templates in an AEM site? (Choose the best answer.)

Answers:

A) Restrict the page creation ability of authors

B) Make effective use of the paragraph system

C) Build more page components

D) Restrict permissions to the templates node

Solution: B

Explanation:

Explanation: Adobe AEM comes with an out-of-the-box component called "Column Control". Using ideas from this component in conjunction with the Paragraph System (the drag-and-drop feature of Adobe AEM) gives content authors the ability to divide and assemble a page in countless combinations. References: https://adobeaemtherightway.wordpress.com/2014/04/08/how-to-reduce-the-number-of-templates- in-your-adobe-aem-project/

Question: 10 *One Answer Is Right*

How can a site be personalized by using social profile data? (Choose the best answer.)

Answers:

A) By using social profile data the credit card numbers will be delivered to the marketer automatically.

B) By using social profile data the website will get a better Google Page Ranking.

C) By using social profile data the performance of the website can be increased from a system administration point of view.

D) Based on the user attributes the marketer is able to show only relevant products for the user within a webshop.

Solution: D

Chapter 6: 9A0-389 - Adobe Campaign Developer

Exam Guide

Adobe Campaign Developer 9A0-389 Exam:

Adobe Campaign Developer 9A0-389 Exam is a certification exam for Adobe Experience Manager Campaign Developer. This exam is related to the Adobe Certified Expert certification. 9A0-389 Exam validates the ability to configure and extend campaign data model, troubleshoot and diagnose related to campaign implementation, demonstrate the ability to use XML to build forms based on the given schema. It also deals with the ability to explain how to efficiently configure predefined filters and demonstrate the ability to hide information from non-privileged users.

9A0-389 Exam topics:

Candidates must know the exam topics before they start of preparation. Because it will really help them in hitting the core. Our **9A0-389 dumps** will include the following topics:

- System Installation and Configuration: 15%
- Customization: 40%
- Operator Security: 7%
- Marketing Campaign Setup: 23%
- Maintenance: 15%

Certification Path:

The Adobe Campaign Developer 9A0-389 Exam certification path includes only one 9A0-389 certification exam.

Who should take the 9A0-389 exam:

The Adobe Certified Expert - AEM Developer certification is an internationally-recognized validation that identifies persons who earn it as possessing skilled in Adobe Experience Manager. If a candidate wants significant improvement in career growth needs enhanced knowledge, skills, and talents. The Adobe Campaign Developer 9A0-389 Exam certification provides proof of this advanced knowledge and skill. If a person has the following necessary knowledge, skills required of an Adobe Campaign Developer 9A0-389 Exam then he should take this exam.

- Design and modify the new schema, including
- Manage the Update-Database Structure wizard
- Increase an existing schema
- Demonstrate schema details (joins, primary keys, auto primary keys, database indexes, enumerations)
- Design and use enumerations
- Change existing forms
- Create new form options
- Create a new external account
- learn architecture models
- Design a data-loading workflow
- Build technical workflow
- Utilize the core workflow activities (including enrichment; core; etc.)
- Package installation
- Understanding of package management

Who should take the 9A0-389 exam:

The Adobe Certified Expert - AEM Developer certification is an internationally-recognized validation that identifies persons who earn it as possessing skilled in Adobe Experience Manager. If a candidate wants significant improvement in career growth needs enhanced knowledge, skills, and talents. The Adobe Campaign Developer 9A0-389 Exam certification provides proof of this advanced knowledge and skill. If a person has the following necessary knowledge, skills required of an Adobe Campaign Developer 9A0-389 Exam then he should take this exam.

- Design and modify the new schema, including
- Manage the Update-Database Structure wizard
- Increase an existing schema
- Demonstrate schema details (joins, primary keys, auto primary keys, database indexes, enumerations)
- Design and use enumerations
- Change existing forms
- Create new form options
- Create a new external account
- learn architecture models
- Design a data-loading workflow
- Build technical workflow
- Utilize the core workflow activities (including enrichment; core; etc.)
- Package installation
- Understanding of package management

How much 9A0-389 Exam Cost:

The price of the 9A0-389 exam is $180 USD.

How to book the 9A0-389 Exam:

These are following steps for registering the 9A0-389 exam.

Step 1: Visit to Pearson Exam Registration

Step 2: Signup/Login to Pearson VUE account

Step 3: Search for Adobe 9A0-389 exam Certifications Exam

Step 4: Select Date, time and confirm with payment method

What is the duration of the 9A0-389 Exam:

Format: Multiple choices, multiple answers

Length of Examination: 80 minutes

Number of Questions: 60

Passing Score: 63%

The benefit in Obtaining the 9A0-388 Exam Certification:

- Adobe Campaign Developer 9A0-389 Exam certification is the absolute best way to communicate your proficiency in the Adobe Experience Manager.
- Resumes with Adobe Campaign Developer certifications get noticed and fast-tracked by hiring managers.
- Adobe Certified Expert - Adobe Campaign Developer recognition and respect from colleagues and employers.
- Adobe Certified Expert - Adobe Campaign Developer receives exclusive updates on Adobe's latest products and innovations.
- Adobe Certified Expert - Adobe Campaign Developer can display Adobe certification logos on their business cards, resumes, and websites to leverage the power of the Adobe brand.

Difficulty in writing 9A0-389 Exam:

Candidates face many problems when they start preparing for the 9A0-389 exam. If a candidate wants to prepare his for the 9A0-389 exam without any problem and get good grades in the exam. Then they have to choose the best **9A0-389 dumps** for real exam questions practice. There are many websites that are offering the latest 9A0-389 exam questions and answers but these questions are not verified by Adobe certified experts and that's why many are failed in their just first attempt. Certification-questions is the best platform which provides the candidate with the necessary 9A0-389 questions that will help him to pass the 9A0-389 exam on the first time. Candidate will not have to take the 9A0-389 exam twice because with the help of **Adobe 9A0-389 exam dumps** Candidate will have every valuable material required to pass the Adobe 9A0-389 exam. We are providing the latest and actual questions and that is the reason why this is the one that he needs to use and there are no chances to fail when a candidate will have valid braindumps from Certification-questions. We have the guarantee that the questions that we have will be the ones that will pass candidate in the 9A0-389 exam in the very first attempt.

For more info visit::

9A0-389 Exam Reference
Adobe Experience Manger Help
Adobe Experience Manger Authoring documentation

Sample Practice Test for 9A0-389

Question: 1 *One Answer Is Right*

How can you give access to a view without giving access to its source folder?

Answers:

A) Provide only read access to the specific parent node.

B) Do not give read access on the parent node of the source folder.

C) Set the "Restrict to information found in sub-folders of" tab to the specific view.

D) The user must be able to access the source folder to access a view.

Solution: B

Explanation:

Explanation: Reference:
https://docs.campaign.adobe.com/doc/archives/en/610/platform-v6.1-en.pdf (page 211)

Question: 2 *One Answer Is Right*

The following XML expression checks for the existence of a first name and will raise an error if the first name field is empty.

```
<leave>
    <check expr="@firstName!=''">
        <error>Please enter a first name!</error>
    </check>
</leave>
```

Where should this expression be located?

Answers:

A) Inside the

element of a form only

B) Inside the master container element within the form

C) Inside the main table element of the data schema.

D) Inside an external JavaScript file which is linked to from within a form container.

Solution: C

Explanation:

Explanation: Reference:
https://docs.campaign.adobe.com/doc/archives/en/610/config uration-v6.1-en.pdf (p.75)

Question: 3 *One Answer Is Right*

In order to restrict the choice of target elements in an input form a sysfilter was added via the link definition:

```
1 <input xpath="mainContact">
2        <!-- Filter the selection of the link on the adobe.com domain -->
3              <sysFilter>
4                   ...
5              </sysFilter>
6 </input>
```

What is missing in line 4?

Answers:

A)

B)

C)

D)

Solution: B

Explanation:

Explanation: Reference:
https://docs.campaign.adobe.com/doc/AC6.1/en/CFG_Input_fo
rms_Form_structure.html

Question: 4 *One Answer Is Right*

There are two distinct recipient folders for France and Germany.
There are two operator groups for France and Germany that are
assigned to each respective folder. A new regional manager
joins the company and needs to be able to see both German and
French recipients. How should you setup the rights of that user?

Answers:

A) Assign the new operator to the Administrator group.

B) Assign the new operator to both French and German folders.

C) Assign the new operator to both French operator and the
German operator group.

D) Assign the new operator to a new operator that has rights on
both folders.

Solution: A

Question: 5 *One Answer Is Right*

What is the most problematic issue about the workflow shown
below?

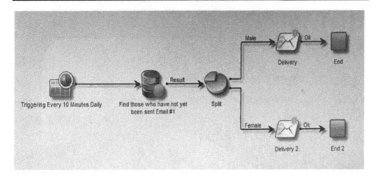

Answers:

A) The end activity will cause all records in the context to be removed from memory.

B) The split might fail causing no deliveries to be sent.

C) The two deliveries will be sent concurrently causing issues with the MTA.

D) The query may not have been completed before the scheduler runs again.

Solution: D

Question: 6 *One Answer Is Right*

Which is required when creating a new Plan?

Answers:

A) Description

B) Parent

C) Start and end date

D) Nature

Solution: D

Explanation:

Explanation: Reference:
https://docs.campaign.adobe.com/doc/AC6.1/en/
CMP_Marketing_campaigns_Setting_up_marketing_campaigns.ht
ml

Question: 7 *One Answer Is Right*

A user receives an Adobe Campaign email notification where they are informed that approval is needed for a certain delivery. The user logs into the Adobe Campaign console and wants to approve the mentioned delivery via the delivery dashboard but the approval link is NOT displayed there. What would cause the approval link to NOT be shown?

Answers:

A) The delivery has already been approved by another operator.

B) The user needs to belong to the "Delivery operators" group in order to see the link.

C) Only the link provided in the email notification can be used to approve the delivery.

D) The content of the delivery needs to be approved by another operator first.

Solution: A

Question: 8 *One Answer Is Right*

What is the purpose of the "Taken into account if" expression on a predefined filer?

Answers:

A) The "Taken into account if" expression provides the marketing user with a drop-down to select which conditions to apply.

B) The conditions on which the "Taken into account if" expression is applied will only be used if the "Taken into account if" expression is true.

C) The "Taken into account if" expression is used to only display an input when the expression is true.

D) The "Taken into account if" expression can be used to format the filter result

Solution: D

Explanation:

Explanation: Reference:
https://docs.campaign.adobe.com/doc/archives/en/610/platfo
rm-v6.1-en.pdf (p.170)

Question: 9 *One Answer Is Right*

You want to add the following new columns to the output of a direct mail file extract: - Firstparturl is already declared in a JavaScrips activity which will be executed before the targeting query: vars.firstparturl =
'http://www.amazingcompany.com/u='; The additional column for recipient called John Doe should look like this:
http://www.amazingcompany.com/u=John-Doe How would you have created the expression in the query activity (Additional columns window) in order to meet the above mentioned requirement?

Answers:

A) 'vars.firstparturl'+vars.firstName+'-'+vars.lastName

B) $(vars/@firstparturl)+@firstName+'-'+@lastName

C) $(vars.firstparturl)+firstName+'-'+lastName

D) 'vars.firstparturl'+$(vars/@firstName)+'-'+$(vars/@lastName)

Solution: D

Question: 10 *One Answer Is Right*

You want to see the email column in the list of the field recipient records. What should you do?

Answers:

A) Add the email to the navtree definition.

B) Add the email to the form definition.

C) Add the email to the "configure list" menu.

D) Add the email to the schema definition.

Solution: C

Chapter 7: 9A0-395 - Adobe Campaign Business Practitioner

Exam Guide

Adobe Campaign Business Practitioner 9A0-395 Exam:

Adobe Campaign Business Practitioner 9A0-395 Exam which is a requisite to obtain Campaign Business Practitioner Certification. This exam measures the Candidates ability and knowledge in Personalize a delivery by using simple attributes, send a proof of the delivery, format an extraction file for direct mail and implement Campaign Tactics in the WorkFlow for example identifying responders and identifying buyers Database Administrator, Digital Marketing Manager and AC Business Practitioner usually hold or pursue this certification and you can expect the same job role after completion of this certification.

9A0-395 Exam topics:

Candidates must know the exam topics before they start of preparation. Because it will really help them in hitting the core. Our **9A0-395 dumps** will include the following topics:

- Installing accessing and navigating the Adobe Campaign Client 11%
- Planning marketing programs 13%
- Segmenting campaign audiences 22%
- Devising a contact strategy 13%

- Configuring a delivery 17%
- Executing and monitoring a campaign 11%
- Measuring campaign results 13%

Certification Path:

The Adobe Campaign Business Practitioner certification path includes only one 9A0-395 certification exam.

Who should take the 9A0-395 exam:

The Adobe Campaign Business Practitioner 9A0-395 Exam certification is an internationally-recognized validation that identifies persons who earn it as possessing skilled as Campaign Business Practitioner. If a candidate wants significant improvement in career growth needs enhanced knowledge, skills, and talents. The Adobe Campaign Business Practitioner 9A0-395 Exam certification provides proof of this advanced knowledge and skill. If a candidate has knowledge of associated technologies and skills that are required to pass Adobe Campaign Business Practitioner 9A0-395 Exam then he should take this exam.

How to study the 9A0-395 Exam:

There are two main types of resources for preparation of certification exams first there are the study guides and the books that are detailed and suitable for building knowledge from ground up then there are video tutorial and lectures that can somehow ease the pain of through study and are comparatively less boring for some candidates yet these demand time and concentration from the learner. Smart Candidates who want to build a solid foundation in all exam topics and related technologies usually combine video lectures with study guides to reap the benefits of both but there is one crucial preparation tool as often overlooked by most candidates the practice exams. Practice exams are built to make students comfortable with the real exam environment. Statistics have shown that most students fail not due to that preparation but due to exam anxiety the fear of the unknown. Certification-questions.com expert team recommends you to prepare some notes on these topics along with it don't forget to practice 9A0-395 Exam dumps which been written by our expert team, Both these will help you a lot to clear this exam with good marks.

How much 9A0-395 Exam Cost:

The price of 9A0-395 exam is $180 USD.

How to book the 9A0-395 Exam:

These are following steps for registering the 9A0-395 exam.
Step 1: Visit to Pearson Exam Registration

- Step 2: Signup/Login to Pearson VUE account
- Step 3: Search for Adobe 9A0-395 Exam Certifications Exam
- Step 4: Select Date, time and confirm with payment method

   What is the duration of the 9A0-395 Exam:

- Format: Multiple choices, multiple answers
- Length of Examination: 75 minutes
- Number of Questions: 46
- Passing Score: 63%

   The benefit in Obtaining the 9A0-395 Exam Certification:

- Resumes with Adobe Certified Campaign Business Practitioner certifications get noticed and fast-tracked by hiring managers.

- Adobe Certified Campaign Business Practitioner recognition and respect from colleagues and employers.
- Adobe Certified Campaign Business Practitioner receives exclusive updates on Adobe's latest products and innovations.
- Adobe Certified Campaign Business Practitioner can display Adobe certification logos on their business cards, resumes, and websites to leverage the power of the Adobe brand.

Difficulty in writing 9A0-395 Exam:

Mostly job holder candidates give a short time to their study and want to pass the exam with good marks. Thereby we have many ways to prepare and practice for exams in a very short time that help the candidates to ready for exams in a very short time without any tension. Candidates can easily prepare Adobe 9A0-395 exams from Certification-questions because we are providing the best **9A0-395 exam dumps** which are verified by our experts. Certification-questions has always verified and updated 9A0-395 dumps that helps the candidate to prepare his exam with little effort in a very short time. We also provide latest and relevant study guide material which is very useful for a candidate to prepare easily for Adobe 9A0-395 exam dumps. Candidate can download and read the latest dumps in PDF and VCE format. Certification-questions is providing real questions of **9A0-395 dumps**. We are very fully aware of the importance of student time and money that's why Certification-questions give the candidate the most astounding brain dumps having all the inquiries answer outlined and verified by our experts.

For more info visit::

9A0-395 Exam Reference

Sample Practice Test for 9A0-395

Question: 1 *One Answer Is Right*

A Business Practitioner is starting delivery for two emails. Which status and failure type combinations are possible in the Recipient Delivery logs for the two messages?

Answers:

A) Status: Sent. Failure Type: Not defined Status: Pending. Failure Type: Not defined

B) Status: Sent. Failure Type: Not defined Status: Pending. Failure Type: Unreachable

C) Status: Sent. Failure Type: Mailbox full Status: Pending. Failure Type: Unreachable

D) Status: Sent. Failure Type: Mailbox full Status: Pending. Failure Type: Not defined

Solution: C

Question: 2 *One Answer Is Right*

A campaign has been setup with an operator group (with 2 operators) as approvers in the "Approvals" section of the delivery. What will happen if one of the operators is unavailable to provide approval?

Answers:

A) The input of just the first approver is required in order for the process to continue onto the next activity

B) The campaign process will not proceed if both the first and seconds approvers do not provide inputs before expiry

C) The process will continue without approval if one has been provided before expiry

D) The process waits indefinitely for inputs from both approvers before moving to the next activity

Solution: B

Question: 3 *One Answer Is Right*

In an email delivery, the Business Practitioner uses conditioned content in the creative that varies by recipient segment. Which three methods allow the Business Practitioner to send proofs that can be set up to cover all variations in the content? (Choose three.)

Answers:

A) Proof with defined proof target

B) Proof with Seed address

C) Proof with Substitution of address with a random profile

D) Proof with Substitution of address with a fixed profile

Solution: A, B, C

Question: 4 *One Answer Is Right*

A large target population spans multiple age groups. The Business Practitioner needs to create a control group with 10% from each age group. What is the most efficient method to create the control group?

Answers:

A) Use a query activity to query the target population and another query activity to obtain the desired control group. Then use an exclusion activity to get the exclusive target.

B) Modify the target population query to not include the control group.

C) Use a query activity to obtain the target population and send to a split activity to remove each age group with record count limitation set as random.

D) Use the out of the box control group functionality with Random Sampling and Data Grouping.

Solution: C

Question: 5 *One Answer Is Right*

A Practitioner chooses the wrong parent folder for the Marketing Plan. How should the Practitioner correct the mistake?

Answers:

A) delete the Marketing Plan and start over

B) edit the plan properties and select a different parent folder

C) drag and drop the Marketing Plan to a new folder

D) rename the parent folder

Solution: C

Question: 6 *One Answer Is Right*

What are two uses of an Exclusion activity in a campaign workflow? (Choose two.)

Answers:

A) to present exclusive offers in an upsell campaign

B) to obtain an accurate population count before Delivery

C) to suppress Recipient on a prior contact list

D) to prioritize contacts with the campaign target

Solution: C, D

Explanation:

Explanation: Reference:
https://helpx.adobe.com/campaign/standard/automating/usin
g/exclusion.html

Question: 7 *One Answer Is Right*

An email and direct mail is sent with the below recipient as the
target. Their blacklist is shown below.

	blackList	blackListEmail	blackListPostalMail
Recipient A	True	True	False
Recipient B	True	False	False
Recipient C	False	True	False
Recipient D	False	False	False

Which recipients receive deliveries and by which channel?

Answers:

A) Email: B,D; Direct Mail: A,B,C,D

B) Email: A,C; Direct Mail:

C) Email: D; Direct Mail: C,D

D) Email: A,B,C; Direct Mail: A,B

Solution: A

Question: 8 *One Answer Is Right*

What happens to a folder when the option 'This folder is a view' is selected?

Answers:

A) The folder shows all items in the folder's sub folders irrespective of item type.

B) The folder shows items that have been deleted in the folder.

C) The folder shows all items of different types irrespective of which folder the items are located in.

D) The folder shows all items of the folders' type irrespective of which folder the items are located in.

Solution: C

Explanation:

Explanation: Reference:
https://forums.adobe.com/thread/2464438

Question: 9 *One Answer Is Right*

The entire list of recipients for a campaign is unavailable at the start of an A/B test. The final list of the recipients will be placed in the finalRecipients folder but they will be loaded by the time the A/B test is complete. To start the A/B test, a list of exclusive recipients is used for the initial deliveries in the folder testRecipients. How should the Business Practitioner modify the workflow diagram?

Answers:

A) modify the existing query to query all recipients in folders 'testRecipients' and 'finalRecipients'

B) modify the original query to all recipients in 'testRecipients' and modify the split activity to query the folder 'finalRecipients' in the complement

C) add the enrichment after the wait to add all recipients in the 'finalRecipients' folder to the complement from the split activity

D) modify the original query to all recipients in 'testRecipients': change the A & B branches to 50% each, and add a new query after the wait to get all recipients in the 'finalRecipients' folder

Solution: B

Question: 10 *One Answer Is Right*

A query on recipients is added to a workflow. The query needs to be able to obtain the last three transactions for each recipient. Which method should be used to obtain this result?

Answers:

A) In the complementary information section, add data of the type 'Data linked to the filtering dimension'.

B) In the complementary information section, add data of the type 'Data of the filtering dimension'.

C) In the advanced tab, add an Initialization Script.

D) Switch the targeting and filtering dimension and add aggregate columns to get the transactions.

Solution: B

Explanation:

Explanation: Reference:
https://docs.campaign.adobe.com/doc/AC/en/
WKF_Repository_of_activities_Targeting_activities.html

Chapter 8: 9A0-397 - Adobe Experience Manager DevOps Engineer

Exam Guide

Adobe Experience Manager DevOps Engineer 9A0-397 Exam:

Adobe Experience Manager DevOps Engineer 9A0-397 Exam is a certification exam for Adobe Experience Manager DevOps Engineer. This exam is related to the Adobe Certified Expert certification. 9A0-397 Exam validates candidate's knowledge of installing platform instances, Configuring AEM, maintaining & optimizing AEM, administrating application operations and troubleshooting AEM. Adobe AEM Dev/Ops Engineers usually hold or pursue this certification and you can expect the same job roles after completion of this certification.

9A0-397 Exam topics:

Candidates must know the exam topics before they start of preparation. Because it will really help them in hitting the core. Our **9A0-397 dumps** will include the following topics:

*   Installing platform instances: 10%
*   Configuring AEM: 16%
*   Maintaining and optimizing AEM: 26%
*   Administrating application operations: 22%

- Troubleshooting AEM: 26%

    Certification Path:

The Adobe Experience Manager DevOps Engineer 9A0-397 Exam certification path includes only one 9A0-397 certification exam.

Who should take the 9A0-397 exam:

The Adobe Certified Expert - AEM DevOps Engineer certification is an internationally-recognized validation that identifies persons who earn it as possessing skilled in Adobe Experience Manager. If a candidate wants significant improvement in career growth needs enhanced knowledge, skills, and talents. The Adobe Experience Manager DevOps Engineer 9A0-397 Exam certification provides proof of this advanced knowledge and skill. If a person has the following necessary knowledge, skills required of an Adobe Experience Manager DevOps Engineer 9A0-397 Exam then he should take this exam.

- Recognize which AEM runtime parameters need to be monitored
- Investigate log data
- Handling initial interpretation of heap and thread dumps
- Configure the performance of hardware, operating system, and the JVM (not coding optimization)
- Tuning a specific Out-of-box AEM
- Monitoring out-of-box AEM metrics
- Determine custom metrics for an installation
- Control AEM to meet custom performance metrics
- Determine measurement points required to monitor customer-specific code
- Describe runtime dependencies and monitoring practices for third-party integrations

- Control, setup, and monitor maintenance processes
- Design and revise disaster recovery plans
- Design tools for debugging AEM, configuring AEM, etc
- Implement tasks from the security checklist to the environment
- Set security checklist to meet environment-specific needs

How to study the 9A0-397 Exam:

Certification-questions.com expert team recommends you to prepare some notes on these topics along with it don't forget to practice **Adobe Experience Manager DevOps Engineer 9A0-397 Dumps** which been written by our expert team, Both these will help you a lot to clear this exam with good marks.

- implement the appropriate procedure to configure the AEM instance as an author or publish
- implement the procedures necessary to configure the AEM instance to use MongoDB, TarMK or other persistence layers to develop a testing roadmap
- implement the procedure to install Apache and dispatcher module
- implement the procedure used to set the run mode
- implement the procedure to configure the replication agents
- implement the procedure to configure custom loggers
- implement the security checklist
- implement the procedure to configure global AEM OSGi settings
- implement the procedure to configure the AEM dispatcher
- implement the procedure to backup and restore AEM
- define when to use online compaction or offline compaction

- implement the procedure to configure regularly scheduled maintenance
- investigate specific storage usage by the application
- implement different procedures to configure JVM runtime settings
- implement the procedure to take a publish instance out of production
- implement the procedure to deploy a package or bundle
- implement best practices to set up a continuous deployment job
- implement the procedure to install an AEM Hotfix or Feature Pack to an existing AEM instance (Author and/or Publish)
- define the cause of HTTP output issues to resolve the issue
- define the cause of performance problems to resolve the issue
- assess the monitoring system for the AEM environment (e.g., dispatcher, AEM instances, the operating system, disk utilization, application metrics)
- define and resolve the rights management issues
- define and resolve the configuration and deployment issues
- implement the procedures to clone the AEM environment

How much 9A0-397 Exam Cost:

The price of the 9A0-397 exam is $180 USD.

How to book the 9A0-397 Exam:

These are following steps for registering the 9A0-397 exam.
Step 1: Visit to Pearson Exam Registration
Step 2: Signup/Login to Pearson VUE account

Step 3: Search for Adobe 9A0-397 exam Certifications Exam
Step 4: Select Date, time and confirm with payment method

What is the duration of the 9A0-397 Exam:

Format: Multiple choices, multiple answers
Length of Examination: 75 minutes
Number of Questions: 50
Passing Score: 63%

The benefit in Obtaining the 9A0-397 Exam Certification:

* Adobe Experience Manager DevOps Engineer 9A0-397 Exam certification is the absolute best way to communicate your proficiency in the Adobe Experience Manager.
* Resumes with Adobe Experience Manager DevOps Engineer certifications get noticed and fast-tracked by hiring managers.
* Adobe Certified Expert - Adobe Experience Manager DevOps recognition and respect from colleagues and employers.
* Adobe Certified Expert - Adobe Experience Manager DevOps Engineer receives exclusive updates on Adobe's latest products and innovations.
* Adobe Certified Expert - Adobe Experience Manager DevOps Engineer can display Adobe certification logos on their business cards, resumes, and websites to leverage the power of the Adobe brand.

    Difficulty in writing 9A0-389 Exam:

Adobe Experience Manager DevOps Engineer 9A0-397 certification exam has a higher rank in the IT sector. Candidate can add most powerful 9A0-397 certification on their resume by passing Adobe 9A0-397 exam. 9A0-397 is a very challenging exam Candidate will have to work hard to pass this exam. With the help of Certification-questions provided the right focus and preparation material passing this exam is an achievable goal. Certification-Questions provide the most relevant and updated **Adobe 9A0-397 exam dumps**. Furthermore, We also provide the **9A0-397 practice test** that will be much beneficial in the preparation. Our aims to provide the best **Adobe 9A0-397 pdf dumps**. We are providing all useful preparation materials such as **9A0-397 dumps** that had been verified by the Adobe experts, **9A0-384 braindumps** and customer care service in case of any problem. These are things are very helpful in passing the exam with good grades.

For more info visit::

9A0-397 Exam Reference
Adobe Experience Manger Help
Adobe Experience Manger Authoring documentation

Sample Practice Test for 9A0-397

Question: 1 *One Answer Is Right*

Multiple OSGi configurations exist in multiple locations within the Java Content Repository (JCR). In which order do these configurations apply?

Answers:

A) configurations with the prod runmode and then /etc/config./apps

B) /configurations with the most matching runmodes and then /apps/confisg, /jcr/configs

C) configurations with the prod runmode and then /libs/configs, /apps/configs

D) /apps./libs and then configurations with the most matching runmodes

Solution: D

Question: 2 *One Answer Is Right*

When configuring the dispatcher module, what setting is required to make sure the dispatcher is used?

Answers:

A) SetHandler dispatcher-handler

B) SetHandler dispatcher-module

C) SetModule module-dispatcher

D) SetModule handler-dispatcher

Solution: B

Question: 3 *One Answer Is Right*

Which three regular maintenance tasks should be executed and scheduled regularly? (Choose three.)

Answers:

A) Audit log purge

B) User cache purge

C) Workflow purge

D) Version purging

Solution: A, B, C

Question: 4 *One Answer Is Right*

A DevOps engineer needs to install bundles via content packages for a continuous deployment setup. In which two ways should this be configured? (Choose two.)

Answers:

A) The JCR installer will only update the bundle if the filename changes; always use a unique bundle version in the filename

B) For an out-of-the-box installation, bundles can be installed from /apps./libs and /etc

C) To install bundles for a certain runmode only, the folder name can be used to limit environments where a bundle is installed

D) Before the deployment of the new bundle, remove the old bundle from JCR as first deployment step to make sure the bundle is updated

E) Whenever the JCR installer detects a change of a bundle file, it will install it to the Felix runtime

F) The JCR installer detects bundle changes at arbitrary depths in the JCR tree and automatically deploys those changes bundles

Solution: A, B

Question: 5 *One Answer Is Right*

A DevOps engineer needs to install a package to a remote AEM staging system using continuous deployment. Which two tools and methods should the DevOps engineer use? (Choose two.)

Answers:

A) curl based script with package manager REST API

B) make file with aem-deploy plugin

C) content package maven plugin

D) apache ant aem-deploy plugin

Solution: B, C

Question: 6 *One Answer Is Right*

A DevOps engineer checks the OSGI console after deployment of a bundle. The OSGI bundle is in the installed state and is not starting. Where should the DevOps engineer look for issues?

Answers:

A) The install log

B) The bundle install log within the OSGI console

C) The error log

D) The OSGI install log within the CRX

Solution: A

Question: 7 *One Answer Is Right*

From which three locations can CRX packages be installed using an out-of-the-box configuration? (Choose three.)

Answers:

A) /apps

B) /content/dam

C) /home/groups

D) /var

E) /libs/system

F) /etc/packages

Solution: B, E, F

Question: 8 *One Answer Is Right*

When running an AEM instance in Production Ready Mode, which two security measures are applied? (Choose two.)

Answers:

A) WebDAV Access to repositories will only be available on author instances.

B) Author user account is disabled on publish instances.

C) The CRXDE Support bundle (com.adobe.granite.crxde-support) is disabled.

D) HTTPS transport layer is enabled.

Solution: A, D

Explanation:

Explanation: Reference https://helpx.adobe.com/experience-manager/6-3/sites/administering/using/security-checklist.html

Question: 9 *One Answer Is Right*

How should a DevOps engineer increase offline tar compaction performance?

Answers:

A) Set "> /dev/nul 2>&1"

B) Increase memory allocated to the Java container

C) Set -Dcompaction-progress-log to 1

D) Set -Dtar.memoryMapped to true

Solution: D

Explanation:

Explanation: Reference http://www.aemcq5tutorials.com/tutorials/online-offline-tar-compaction-in-aem/ #performance_tar_compact

Question: 10 *One Answer Is Right*

How should the run mode be configured to set it to "UAT"?

Answers:

A) Rename the quickstart.jar filename to contain the run mode before the initial UAT installation.

B) Set the run mode in the system properties in the start script on all systems of the UAT environment and restart the system.

C) Configure the run mode on the Felix console of all systems in the UAT environment.

D) Create a configuration package that contains this run mode, deploy it on the author of the UAT environment, and replicate the package.

Solution: B

Chapter 9: 9A0-409 - Adobe Premiere Pro CC 2015 ACE

Exam Guide

How to ready for 9A0-409 Certification Exam

For 9A0-409 Certification Exam here is our preparation guide

9A0-409: Get our instant guide if you don't have time to read all the page

The ADOBE 9A0-409 certification exam is of paramount importance both in professional life and in the ADOBE certification process. With ADOBE certification, you can easily get a good job in the market and move towards success. Specialists who have successfully finished the ADOBE 9A0-409 exam preparation are the absolute favorites in the industry. You will pass the ADOBE 9A0-409 certification exam and have career chances.

The 9A0-409 is well recognized as ACE: Premiere Pro CC 2015 , like all examinations, ADOBE has some freedom to examine a variety of topics. This means that most of the contents of 9A0-409 are necessary as they perform random tests on the many available topics. Also keep in mind that experience conditions often exist because they observed the average person and what is needed. You can always go further to succeed with the 9A0-409, but it can take some extra effort

In this advanced age, getting an excellent Adobe certification exam has become more necessary for Adobe ACE specialists. If we examine the world of IT credentials, we will find many certified exams, but the truth is that Adobe 9A0-409 certification is above all the credentials available in the IT profession. The Adobe Premiere Pro CC 2015 ACE certification is one of the best ways to increase value in the IT world. You want to know why? The 9A0-409 exam module was recently presented by Adobe and has attracted the attention of many Adobe ACE professionals and aspiring people who wish to increase their credibility in the market.

Staying focused on studying can be difficult, but take it in mind that the best jobs in the world are only several tests away. Regardless of whether you enter cybersecurity or do a basic level technical job, certification is a clear, learning and rewarding path for careers that pay a LOT of money. They offer a better balance between professional and private life and can get in touch with some of the leaders in the business world.

The exam objectives are strange for each exam and are usually provided by the certification provider. These normally indicate to the candidate which subjects are relevant, what he should know and why the exam tries to cover these subjects. It is necessary to find them out for your precise exam. This can be found on almost all provider websites and reports a lot like studies.

Adobe 9A0-409 Exam Intro:

The Adobe Certified Expert (ACE) certification is the industry-approved validation of your skills in Adobe Premiere Pro CC. This certification requires an understanding of the design elements in-depth, when making videos, perfecting the visual and audio parts of a project, editing video sequences, furthermore experience with Adobe Media Encoder and editing in a professional environment.

The topics of 9A0-409 Exam:

Candidates must know the exam topics before they start of preparation. Because it will really help them in hitting the core. Our **9A0-409 dumps** will include the following topics:

- Import, Organize, and Select Assets
- Work with Media and Sequence
- Adjust the Audio Properties of Clips and Sequences
- Use Effects, Transitions and Titles
- Exchange and Manage Media and Projects

Here are the requirements of 9A0-409 Exam:

There is no prerequisite for ADOBE 9A0-409 certification.

9A0-409 Exam Format:

In order to apply for the 9A0-409, You have to follow these steps

1. Go to the 9A0-409 Official Site
2. Read the instruction Carefully
3. Follow the given steps
4. Apply for the 9A0-409

Here is a procedure to apply for 9A0-409 Exam:

Format: Multiple choices, multiple answers

- Length of Examination: 90 minutes
- Number of Questions: 60
- Passing score: 300 to 700
- Language: English

Here is a salary of 9A0-409 Certified Professional:

- United States: 71,000 USD
- India: 64,110 INR
- Europe: 64,396 Euro
- England: 54,305 Pound

Here is the price of 9A0-409 Exam:

The price of 9A0-409 exam is $180 USD.

Here are the advantages of Obtaining the 9A0-409 Certification Exam:

- Adobe Certified Expert are distinguished among competitors. Adobe Certified Expert certification can give them an edge at that time easily when candidates appear for a job interview employers seek to notify something which differentiates the individual to another.
- Adobe Certified Expert have more useful and relevant networks that help them in setting career goals for themselves. Adobe Certified Expert networks provide them with the right career direction than non certified usually are unable to get.
- Adobe Certified Expert will be confident and stand different from others as their skills are more trained than non-certified professionals.
- Adobe Certified Expert have the knowledge to use the tools to complete the task efficiently and cost effectively than the other non-certified professionals lack in doing so.
- Adobe Certified Expert Certification provides practical experience to candidates from all the aspects to be a proficient worker in the organization.
- Adobe Certified Expert Certifications provide opportunities to get a job easily in which they are

interested in instead of wasting years and ending without getting any experience.

Here are the Difficulty in taking the 9A0-409 Exam:

All Candidates wants to get success in the 9A0-409 exam in the just first attempt but mostly not been able to get success in it due to poor selection of their 9A0-409 training material. Certification-questions.com **9A0-409 dumps** are the perfect way to prepare 9A0-409 exam to get good grades in the just first attempt. Certification-questions has quality **9A0-409 dumps** and their ADOBE Certified professionals designed them emphatically than others. Certification-questions is renowned across the globe just because of their quality study material So if candidates want instant success in the 9A0-409 exam with quality 9A0-409 training material then Certification-questions is the best option for you because our management is well trained in it and we update each question of all exams on regular basis after consulting recent updates with their ADOBE certified professionals. It is very easy for the candidates to download 9A0-409 exam dumps pdf from Certification-questions. With the help of **9A0-409 dumps**, candidates will get all the latest questions and answers for 9A0-409 exam. We are confident that candidates can get a high score with excellent grades for the 9A0-409 exam.

Here is the guide to get ready for the 9A0-409 Exam:

Certification-Questions is well recognized for a variety of dumps for 9A0-409 certification. Taking certificates is not an easy job since students have to study carefully.9A0-409 education also takes a long time. Therefore, when considering student needs, we design many landfills for them with 9A0-409 questions. Our products, including the study guide, will help students pass exams.

Each practice exam contains questions and answers which help students pass to their final exam. After taking and understanding our modules, you will pass the exam. But it doesn't stop there; You will always succeed in your profession thanks to our complete guides. In the future, you can make your own products.

For you, we have an advanced way to prepare each material for you. We have used the latest information in the production of each product. Our **9A0-409 dumps** are easy to use, so everyone can understand them.

No one likes failure, mostly in such complex fields where certification requires a lot of research, planning and attention. A single attempt is so tough that it could even break the nerves of the students. Our dumps are so valid and best which will able you to pass your 9A0-409 without any pain.

To get more information visit::

9A0-409 Exam Reference

Chapter 10: 9A0-410 - Adobe Experience Manager Forms Developer ACE Exam

Exam Guide

Adobe Experience Manager Forms Developer 9A0-410 Exam:

Adobe Experience Manager Forms Developer 9A0-410 Exam is related to Adobe Certified Expert Certification. This exam validates the Candidates ability to apply procedural concepts for developing Javascript code for complex form experiences, and to implement form sets, and to build an adaptive form or document, manage XFA templates using Form Manager. It also deals with ability to to determine an appropriate custom submit action given a use case, as well as recognize variables related to the generation of a document of record based upon form submission, to determine appropriate parameters to configure OSGI watched folders for batch generation based on business requirements, determine correspondence management (CM) assets given a use case, diagnose and troubleshoot failures and errors, and apply procedural concepts to implement and configure Forms Portal Adobe AEM Forms Developer Exam Guide 5 component.

9A0-410 Exam topics:

Candidates must know the exam topics before they start of preparation. Because it will really help them in hitting the core. Our **9A0-410 dumps** will include the following topics:

- Complex form implementation 42%
- Form submission handling 18%
- Workflow implementation 20%
- Correspondence management 5%
- Testing and debugging 10%
- Form portal configuration 5%

Certification Path:

The Adobe Experience Manager Forms Developer certification path includes only one 9A0-410 certification exam.

Who should take the 9A0-410 exam:

The Adobe Experience Manager Forms Developer 9A0-410 Exam certification is an internationally-recognized validation that identifies persons who earn it as possessing skilled as Adobe Certified Expert. If a candidate wants significant improvement in career growth needs enhanced knowledge, skills, and talents. The Adobe Experience Manager Forms Developer 9A0-410 Exam certification provides proof of this advanced knowledge and skill. If a candidate has knowledge of associated technologies and skills that are required to pass Adobe Experience Manager Forms Developer 9A0-410 Exam then he should take this exam.

How to study the 9A0-410 Exam:

There are two main types of resources for preparation of certification exams first there are the study guides and the books that are detailed and suitable for building knowledge from ground up then there are video tutorial and lectures that can somehow ease the pain of through study and are comparatively less boring for some candidates yet these demand time and concentration from the learner. Smart Candidates who want to build a solid foundation in all exam topics and related technologies usually combine video lectures with study guides to reap the benefits of both but there is one crucial preparation tool as often overlooked by most candidates the practice exams. Practice exams are built to make students comfortable with the real exam environment. Statistics have shown that most students fail not due to that preparation but due to exam anxiety the fear of the unknown. Certification-questions.com expert team recommends you to prepare some notes on these topics along with it don't forget to practice 9A0-410 Exam dumps which been written by our expert team, Both these will help you a lot to clear this exam with good marks.

How much 9A0-410 Exam Cost:

The price of 9A0-410 exam is $180 USD.

How to book the 9A0-410 Exam:

These are following steps for registering the 9A0-410 exam. Step 1: Visit to Pearson Exam Registration

- Step 2: Signup/Login to Pearson VUE account
- Step 3: Search for Adobe 9A0-410 Exam Certifications Exam
- Step 4: Select Date, time and confirm with payment method

   What is the duration of the 9A0-410 Exam:

- Format: Multiple choices, multiple answers
- Length of Examination: 95 minutes
- Number of Questions: 80

- Passing Score: 63%

  The benefit in Obtaining the 9A0-410 Exam Certification:

- Resumes with Adobe Certified Expert certifications get noticed and fast-tracked by hiring managers.
- Adobe Certified Expert recognition and respect from colleagues and employers.
- Adobe Certified Expert receives exclusive updates on Adobe's latest products and innovations.
- Adobe Certified Expert can display Adobe certification logos on their business cards, resumes, and websites to leverage the power of the Adobe brand.

  Difficulty in writing 9A0-410 Exam:

This 9A0-410 exam is very difficult to prepare. Because it requires all candidate attention with practice. So, if Candidate wants to pass this 9A0-410 exam with good grades then he has to choose the right preparation material. By passing the 9A0-410 exam can make a lot of difference in your career. Many Candidates wants to achieve success in the 9A0-410 exam but they are failing in it. Because of their wrong selection but if the candidate can get valid and latest 9A0-410 study material then he can easily get good grades in the 9A0-410 exam. Certification-questions providing many 9A0-410 exam questions that help the candidate to get success in the 9A0-410 test. Our **9A0-410 exam dumps** specially designed for those who want to get their desired results in the just first attempt. 9A0-410 braindump questions provided by Certification-questions make candidate preparation material more impactful and the best part is that the training material provided by Certification-questions for 9A0-410 exams are designed by our experts in the several fields of the IT industry.

For more info visit::

Adobe Exams Study Guide & Tests   - *by David Mayer*

9A0-410 Exam Reference

Sample Practice Test for 9A0-410

Question: 1 *One Answer Is Right*

How should a developer support mobile devices for the Search and Lister component?

Answers:

A) Within the edit dialog of the search pane, click the Enable Mobile checkbox

B) Within the edit dialog of the list pane, click the Enable Mobile checkbox

C) Build a custom component

D) Do nothing. The component adjusts accordingly

Solution: D

Explanation:

Explanation: Forms Portal Search & Lister component is mobile device friendly and adapts accordingly. All three default views: Grid, Card, Panel relayouts according to the device in which site is opened provided with the fact that web page also adapts. The simple fact is that, Search & Lister is a component only and does not govern page level styling.

Question: 2 *One Answer Is Right*

A client uploads XDPs from Workbench. The client cannot render the forms as HTML5. What should the client do?

Answers:

A) Change the time interval on the FormsReplicationScheduler

B) Change the profile resource super type in the FormsManager AddOn Configuration

C) Change the FormsManager AddOn Configuration to synchronize synchronously

D) Change the time interval on the FormsManager AddOn Configuration

Solution: B

Question: 3 *One Answer Is Right*

What is a reason to use a form fragments when designing templates to be rendered as PDF?

Answers:

A) To fully integrate with analytics

B) To allow multiple forms to use the same fields

C) To improve publisher system performance

D) To make sure fragments are not replicated to a publisher

Solution: B

Explanation:

Explanation: Reference: https://helpx.adobe.com/aem-forms/6/forms-service.html

Question: 4 *One Answer Is Right*

A Watched Folder endpoint throws a coercion error when consuming an XML variable. What is causing this error?

Answers:

A) A Watched Folder can only consume a Document variable

B) A Watched Folder can only consume a List of Documents

C) A Watched Folder can only consume a String variable

D) A Watched Folder can only consume a Map of Documents

Solution: B

Question: 5 *One Answer Is Right*

Instance Manager must be used to remove a row. Which line of code should be used to perform this task?

Answers:

A)
row_name.instanceManager.removeInstance(this.row_name.instanceIndex);

B)
instanceManager.removeInstance(this.row_name.instanceIndex);

C)
row_name.instanceManager.removeRow(row_name.instanceIndex);

D) instanceManager.removeInstance();

Solution: D

Explanation:

Explanation: Reference:
https://help.adobe.com/en_US/livecycle/11.0/DesignerHelp/
WS92d06802c76abadb11a7f71129b8b00825-7fa6.2.html

Question: 6 *One Answer Is Right*

What is an advantage of implementing an HTML5 form set?

Answers:

A) The ability to replicate to a publisher instance

B) The ability to implement form sets without requiring special permissions on the dispatcher

C) The ability to bundle XFA templates together and restrict those templates based on user input

D) The ability to bundle XFA templates together and display those templates based on user input

Solution: D

Explanation:

Explanation: Reference: https://helpx.adobe.com/aem-forms/6/html5-forms/designing-form-template.html

Question: 7 *One Answer Is Right*

You are creating a form in AEM Forms Designer. The visibility of a form field is set to hidden. What is the result?

Answers:

A) The field is removed from the layout on render

B) The field remains at the same X.Y position and is replaced with a blank space

C) The field is removed from the layout

D) The field remains at the same X.Y position and is removed from the layout

Solution: D

Explanation:

Explanation: Reference:
https://help.adobe.com/en_US/AEMForms/6.1/DesignerHelp/WS107c29ade9134a2c583558f12a7dc955d9-8000.2.html

Question: 8 *One Answer Is Right*

An organization needs a reusable function in a design template that receives a numeric value. How should the function be declared?

Answers:

A) function ProcessData(var value)

B) function ProcessData(value)

C) function ProcessData(int value)

D) public void ProcessData(int value)

Solution: C

Question: 9 *One Answer Is Right*

What is a use case for an Adaptive Document?

Answers:

A) To create an interactive web-based representation of a PDF portfolio that can be embedded into a site

B) To capture user information and drive users through a series of screens (dynamic interactive)

C) To create an Adaptive Form that can be saved offline and allow a mobile workforce to continue to work

D) To display a user account balance after a transaction on a site (non-interactive)

Solution: A

Explanation:

Explanation: Reference: https://helpx.adobe.com/aem-forms/6-1/adaptive-document.html

Question: 10 *One Answer Is Right*

What is the purpose of /libs/fd/fm/content/manage.json?

Answers:

A) To return a list of formsets that match a specific criterion

B) To return a list of servlets that match a specific criterion

C) To return a list of forms that match a given criterion

D) To return a list of custom components that match a specific criterion

Solution: C

Explanation:

Explanation: Reference: https://helpx.adobe.com/aem-forms/6/listing-forms-webpage-using-apis.html

Chapter 11: 9A0-412 - Adobe Analytics Business Practitioner

Exam Guide

Adobe Analytics Business Practitioner 9A0-412 Exam:

Adobe Analytics Business Practitioner 9A0-412 Exam is requisite to obtain Analytics Business Practitioner Certification. This exam focuses on individuals who are currently working or have previous work experience with the job responsibilities as Adobe Analytics Business Practitioner. Adobe Analytics Business Practitioner Certification also plays a positive role in Adobe Specialization a key element in the solution Partner Program. This exam measures the Candidate knowledge and skills in conducting a business analysis, reporting and dashboarding and administering and troubleshooting.

9A0-412 Exam topics:

Candidates must know the exam topics before they start of preparation. Because it will really help them in hitting the core. Our **9A0-412 dumps** will include the following topics:

- Conducting a business analysis 33.4%
- Reporting and Dashboarding 23.3%
- Segmenting 20%
- Administering and troubleshooting 23.3%

  Certification Path:

The Adobe Analytics Business Practitioner certification path includes only one 9A0-412 certification exam.

Who should take the 9A0-412 exam:

The Adobe Analytics Business Practitioner 9A0-412 Exam certification is an internationally-recognized validation that identifies persons who earn it as possessing skilled in Adobe Analytics Business Practitioner Certification. If a candidate wants significant improvement in career growth needs enhanced knowledge, skills, and talents. The Adobe Analytics Business Practitioner 9A0-412 Exam certification provides proof of this advanced knowledge and skill. If a candidate has knowledge of associated technologies and skills that are required to pass Adobe Analytics Business Practitioner 9A0-412 Exam then he should take this exam.

How to study the 9A0-412 Exam:

There are two main types of resources for preparation of certification exams first there are the study guides and the books that are detailed and suitable for building knowledge from ground up then there are video tutorial and lectures that can somehow ease the pain of through study and are comparatively less boring for some candidates yet these demand time and concentration from the learner. Smart Candidates who want to build a solid foundation in all exam topics and related technologies usually combine video lectures with study guides to reap the benefits of both but there is one crucial preparation tool as often overlooked by most candidates the practice exams. Practice exams are built to make students comfortable with the real exam environment. Statistics have shown that most students fail not due to that preparation but due to exam anxiety the fear of the unknown. Certification-questions.com expert team recommends you to prepare some notes on these topics along with it don't forget to practice 9A0-

412 Exam dumps which been written by our expert team, Both these will help you a lot to clear this exam with good marks.

How much 9A0-412 Exam Cost:

The price of 9A0-412 exam is $180 USD.

How to book the 9A0-412 Exam:

These are following steps for registering the 9A0-412 exam.
Step 1: Visit to Pearson Exam Registration

- Step 2: Signup/Login to Pearson VUE account
- Step 3: Search for Adobe 9A0-412 Exam Certifications Exam
- Step 4: Select Date, time and confirm with payment method

What is the duration, language, and format of the 9A0-412 Exam:

- Format: Multiple choices, multiple answers
- Length of Examination: 105 minutes
- Number of Questions: 60
- Passing Score: 60%

The benefit in Obtaining the 9A0-412 Exam Certification:

- Resumes with Adobe Analytics Business Practitioner certifications get noticed and fast-tracked by hiring managers.
- Adobe Analytics Business Practitioner recognition and respect from colleagues and employers.
- Adobe Analytics Business Practitioner receives exclusive updates on Adobe's latest products and innovations.

- Adobe Analytics Business Practitioner can display Adobe certification logos on their business cards, resumes, and websites to leverage the power of the Adobe brand.

  Difficulty in writing 9A0-412 Exam:

Adobe Analytics Business Practitioner 9A0-412 Exam certification exam has a higher rank in the IT sector. Candidate can add most powerful 9A0-412 certification on their resume by passing Adobe 9A0-412 exam. 9A0-412 is a very challenging exam Candidate will have to work hard to pass this exam. With the help of Certification-questions provided the right focus and preparation material passing this exam is an achievable goal. Certification-Questions provide the most relevant and updated Adobe 9A0-412 exam dumps. Furthermore, We also provide the 9A0-412 practice test that will be much beneficial in the preparation. Our aims to provide the best Adobe 9A0-412 pdf dumps. We are providing all useful preparation materials such as **9A0-412 dumps** that had been verified by the Adobe experts, 9A0-412 braindumps and customer care service in case of any problem. These are things are very helpful in passing the exam with good grades.

For more info visit::

9A0-412 Exam Reference

Sample Practice Test for 9A0-412

Question: 1 *One Answer Is Right*

Please answer the question using the segment in the exhibit.

Which statement is true regarding the segment if "visit" was changed to "hit"?

Answers:

A) The segment population will decrease

B) The segment population will increase

C) The number of product views will be less than two

D) The number of product views will be greater than two

Solution: D

Question: 2 *One Answer Is Right*

An analyst needs to deliver a Workspace to the members of the executive team on a weekly schedule. The members of this executive group do NOT for Adobe Analytics or the Adobe Marketing Cloud. How should the analyst schedule the weekly delivery of this Workspace to this executive group?

Answers:

A) Create an "Executive" publishing list and send to that group

B) Add the executive's email address directly in the recipient's field and send

C) Create a new "Executive" user group and send to that group

D) Add the executive's email addresses to an existing user group and send

Solution: D

Question: 3 *One Answer Is Right*

While running an Adobe Analytics eVar classification report, the analyst has identified that one of the classification values should not be in this report. In order to remove only this single classification value, what should the analyst do?

Answers:

A) Upload a new classification value of "~empty~" so that the above report value is deleted

B) Upload a new classification with a blank value where the previous value existed

C) Delete the classification value from the above report via the classification manager

D) Delete the entire classification column from the classification manager

Solution: C

Question: 4 *One Answer Is Right*

A marketing organization has ongoing marketing efforts across multiple channels. In the past, the organization utilized the following URL parameter their efforts: mpid: to identify the marketing agency partner afid: for affiliate ID campId: to track the campaign ID itrkid: for internal campaigns offerid: tracks the special offer ID The marketing manager asks the analyst to provide an example URL that the team can replicate. What is valid landing page URL with appropriate values?

A. domain.com/landing-
page/&afID=value&CAMPid=value&itrkId=value&mpid=value&offerid=value
B. domain.com/landing-
page/?cid=123wes?afid=value&campid=value&itrkid=value&mpid=value&offerid=value
C. domain.com/landing-page/mpid=value&afid=value&campid=value&itrkid=value&offerid=value
D. domain.com/landing-page/mpid=value&afid=value&campid=value&itrkid=value&offerid=value

Answers:

A) Option A

B) Option B

C) Option C

D) Option D

Solution: C

Question: 5 *One Answer Is Right*

An analyst creates a Workspace based on two separate segments: A) Product type = electronics B) Marketing channel = affiliate As a follow-up analysis, the analyst wants to identify electronics sales from the affiliate channel over the past 90 days. Which segment should the analyst create to fulfill these requirements?

Answers:

A) A new segment to include segment A OR B. Then within 90 days the orders event count is greater than "0".

B) A new segment to include segment A AND B. Then within 90 days the orders event count is greater than "0". Product type = electronics

C) A new segment to include segment A. Then within 90 days the orders event count is greater than "0".

D) A new segment to include segment B. Then within 90 days the orders event count is greater than "0". Marketing channel = affiliate

Solution: B

Question: 6 *One Answer Is Right*

Per a solution design reference, the following variables are set when a social share occurs: event5 – Social Share eVar7 – Social Share Channel prop7 – Social Share Channel In Adobe Analytics Reports, which report can be created?

Answers:

A) Social Share per Visit by Social Share Channel

B) Top URLs shared

C) Top Products shared

D) Social Shares by Marketing Channel

Solution: A

Question: 7 *One Answer Is Right*

In the image of the funnel shown in the exhibit, which statement is true if the evenue decreases by 50%?

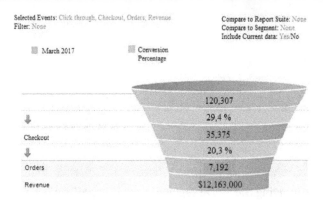

Selected Events: Click through, Checkout, Orders, Revenue
Filter: None

Compare to Report Suite: None
Compare to Segment: None
Include Current data: Yes/No

March 2017   Conversion Percentage

	120,307
⬇	29,4 %
Checkout	35,375
⬇	20,3 %
Orders	7,192
Revenue	$12,163,000

Answers:

A) Average revenue per order increases

B) Average orders per click-through decrease

C) Average revenue per order decreases

D) Average orders per click-through increase

Solution: C

Question: 8 *One Answer Is Right*

An analyst is running the standard "Pages" report and is applying the segment shown in the exhibit.

Title

  Internal Search Content

Description

  Includes the internal search event

Tags

   Add Tags

Definitions

Show   Visit   ▼

  ::   Internal Searches   is greater than or equal to   1

When applying the above segment to the "Pages" report, the analyst sees the following top 5 pages: 1. home 2. cart 3. products:electronics 4. search:computers:1 5. search:televisions:1 What change should be made to remove the non-search result pages from the pages report?

Answers:

A) Run the "Pages" report with the internal search event in Workspace

B) Change the segment container from "Visit" to "Hit"

C) Run the "Pages" report with the internal search event in Reports

D) Change the segment container from "Visit" to "Visitor"

Solution: B

Question: 9 *One Answer Is Right*

An analyst needs to configure an Affiliate marketing channel within the Marketing Channel Processing Rules. Traffic from this channel uses the "source" and "campaign_name" query parameters. The "source" query parameter is used by other marketing channels. Traffic from the Affiliate marketing channel this query parameter with a value that starts with "aff". Affiliate traffic may NOT have a defined value for "campaign name". How should the analyst configure the processing rule so that any hits from affiliate are added to the Affiliate marketing channel?

Answers:

A) Hits where the "source" query parameter starts with "aff" and the "campaign_name" exists

B) Hits where the "source" query parameter starts with "aff"

C) Hits where the "source" and "campaign_name" query parameters exist

D) Hits where the 'campaign_name" query parameter starts with "aff"

Solution: C

Question: 10 *One Answer Is Right*

An analyst needs to share a Workspace with a select group of individuals that may change over time. Which process should the analyst follow?

Answers:

A) Add the Workspace to the user group from the Admin console

B) Share the Workspace with the user group from within the Workspace

C) Select the "Share" option and add each individual that needs to access the project from the Workspace

D) Select the "Share" option, create a new custom group for the set of users and share the project with that group from the Workspace

Solution: A

Chapter 12: AD0-300 - Adobe Campaign Business Practitioner

Exam Guide

How to Prepare For Adobe Campaign Business Practitioner AD0-300 Exam

Preparation Guide for Adobe Campaign Business Practitioner AD0-300 Exam

Introduction

Adobe certification are swift, economical, and official way to demonstrate your skills using a range of industry-standard software packages, so it's probably a good investment. You don't need to have training in these tools before taking the exam, but it is extremely advisable.

Adobe Campaign Business Practitioner AD0-300 Exam:

Adobe Campaign Business Practitioner AD0-300 Exam is related to expertise to help clients understand and value in an Adobe solution.Adobe certification strictly shadows industry accepted process to ensure validity and reliability. This exam validates the ability to translate campaign requirements into an actionable workflow, creating delivery, campaign, control groups, and seed templates, enhancement configuration , readlists creation, controlling user access, manipulation of external and internal data for use in campaign. Use descriptive analysis or query analysis tools.

Adobe Certified Expert Partners, Customers and Consultants usually hold or pursue this certification and you can expect the same job role after completion of this certification.

Topics of Adobe Campaign Standard Developer ACE AD0-300 Exam:

Candidates must know the exam topics before they start of preparation. Because it will really help them in hitting the core. Our **Adobe AD0-300 dumps** will include the following topics:

1. Campaign Management 17%

- Create campaigns, configure campaigns, and determine the correct campaign template.

  2. Workflow Management 33%

- Interpret campaign requirements and setup approvals. Ability to solve workflow errors and determine a correct design for a marketing workflow. Ability to build technical workflows and also execute workflows.

  3. Data Management 15%

- Ability to import data, export data, and perform data investigations. You should also be able to build lists and configure a predefined filter.

  4. Delivery Management 23%

- Creating and configure deliveries. You should also be able to correct proofs for approvals, interpret deliver audits, and deploy a delivery

  5. Reporting 10%

- Generating delivery reports, determining the appropriate report(s) to generate, and interpretation of campaign reports.

6. Administration 2%

- Management of users and folder structures.

Certification Path:

The Adobe Campaign Business Practitioner certification path includes only one AD0-300 certification exam.

Who should take the AD0-300 exam:

The Adobe Campaign Business Practitioner AD0-300 Exam certification is an internationally-recognized validation that identifies persons who earn it posses possessing skilled in Adobe Photoshop. If a candidate wants significant improvement in career growth needs enhanced knowledge, skills, and talents. The Adobe Campaign Business Practitioner AD0-300 Exam certification provides proof of this advanced knowledge and skill. If a candidate has knowledge of associated technologies and skills that are required to pass Adobe Campaign Business Practitioner AD0-300 Exam then he should take this exam.

How to study the AD0-300 Exam:

There are two main types of resources for preparation of certification exams first there are the study guides and the books that are detailed and suitable for building knowledge from ground up then there are video tutorial and lectures that can somehow ease the pain of through study and are comparatively less boring for some candidates yet these demand time and concentration from the learner. Smart Candidates who want to build a solid foundation in all exam topics and related technologies usually combine video lectures with study guides to reap the benefits of both but there is one crucial preparation tool as often overlooked by most candidates the practice exams. Practice exams are built to make students comfortable with the real exam environment. Statistics have shown that most students fail not due to that preparation but due to exam anxiety the fear of the unknown. Certification-questions.com expert team recommends you to prepare some notes on these topics along with it don't forget to practice AD0-300 Exam dumps which been written by our expert team, Both these will help you a lot to clear this exam with good marks.

How much AD0-300 Exam Cost:

The price of AD0-300 exam is $180 USD.

How to book the AD0-300 Exam:

Exams are delivered worldwide by PSI.

These are following steps for registering the AD0-300 exam.

- Step 1: Visit Adobe's credential management system logon page. The logon requires that you have an Adobe ID. If you do not have an Adobe ID, click the link for "Get an Adobe ID"
- Step 2: After logging on, if you have never taken an Adobe exam before, you will be instructed to create a Profile.After creating your Profile, you will be instructed to update your opt-in Settings.
- Step 3: Once logged on, click "Home" then click "Schedule your exam at PSI."
- Step 4: You will be directed to a new page within CertMetrics where you will click "Click here to log in to PSI".
- Step 5: You will be taken to a page hosted by our exam delivery vendor, PSI, that displays the available exams

- Step 6: Scroll through the list of available exams and press the "Schedule Exam" button for the exam you want to take.
- Step 7: Select a delivery mode for your exam by choosing either "Test Center" or "Remote Online Proctored Exam."
- Step 8: Select your exam language to see options for your exam.
- Step 9: Select an available date and start time.
- Step 10: Confirm schedule details to receive a booking confirmation.
- Step 11: Proceed to payment.
- Step 12: After payment is successful, you will receive an email confirmation your registration details and a receipt.

What is the duration, language, and format of AD0-300 Exam:

- Format: Multiple choices, multiple answers
- Length of Examination: 85 minutes
- Number of Questions: 54
- Passing Score: All Adobe exams are reported on a scale of 300 to 700. The passing score for each exam is 550.

The benefit in Obtaining the AD0-300 Exam Certification:

- Resumes with Adobe Certified Expert certifications get noticed and fast-tracked by hiring managers.
- AD0-300 Certification is distinguished among competitors. AD0-300 certification can give them an edge at that time easily when candidates appear for employment interview, employers are very fascinated to note one thing that differentiates the individual from all other candidates.
- AD0-300 certification will be more useful and relevant networks that help them in setting career goals for themselves. AD0-300 networks provide them with the

correct career guidance than non certified generally are unable to get.

- AD0-300 certified candidates will be confident and stand different from others as their skills are more trained than non-certified professionals.

- **Adobe AD0-300 Exam** will provide proven knowledge to use the tools to complete the task efficiently and cost effectively than the other non-certified professionals lack in doing so.

- AD0-300 Certification provides practical experience to candidates from all the aspects to be a proficient worker in the organization.

- AD0-300 Certifications will provide opportunities to get a job easily in which they are interested in instead of wasting years and ending without getting any experience.

- AD0-300 credential delivers higher earning potential and increased promotion opportunities because it shows a good understanding of manage topologies cadences

Difficulty in writing Adobe AD0-300 Exam:

Candidates face many problems when they start preparing for the AD0-300 exam. Certification exams are usually tough and tricky. It is observed that exam dive into minutia, which are usually difficult for professionals even with years of experience. If a candidate wants to prepare his for the AD0-300 exam without any problem and get good grades in the exam.

The candidates have to follow discipline such as organizing study places, taking proper breaks such as for every 1 hour, you study, take a short 10 minute break to recharge. Make studying less overwhelming by condensing notes from course.

Then they have to choose the best AD0-300 dumps for real exam questions practice. There are many websites that are offering the latest AD0-300 exam questions and answers but these questions are not verified by Adobe certified experts and that's why many are failed in their just first attempt. Certification-questions is the best platform which provides the candidate with the necessary AD0-300 questions that will help him to pass the AD0-300 exam on the first time. Candidate will not have to take the AD0-300 exam twice because with the help of **Adobe AD0-300 dumps** Candidate will have every valuable material required to pass the Adobe AD0-300 exam. We are providing the latest and actual questions and that is the reason why this is the one that he needs to use and there are no chances to fail when a candidate will have valid dumps from Certification-questions. We have the guarantee that the questions that we have will be the ones that will pass candidate in the AD0-300 exam in the very first attempt.

For more info visit::

AD0-300 Exam Reference

Sample Practice Test for AD0-300

Question: 1 *One Answer Is Right*

A campaign business practitioner needs to build an A/B test email campaign for two different promotions. By the theme that generates the most website visits. Which email metric indicates the winning promotion theme?

Answers:

A) Highest Click through rate

B) Lowest unsubsubscribe rate

C) Highest open rate

D) Lowest activity rate

Solution: A

Question: 2 *One Answer Is Right*

Which best practice should a user take to move an improperly placed campaign into the correct program folder?

Answers:

A) Rename the improper program location to the appropriate program

B) Recreate the campaign in correct program

C) Select the correct program dropdown in the campaign edit tab

D) Click and drag the campaign into the appropriate program

Solution: C

Explanation:

Explanation: Reference:
https://docs.campaign.adobe.com/doc/AC/en/
PTF_Administration_basics_Access_management.html

Question: 3 *One Answer Is Right*

A user is assigned to the administrator group (which has full access) and a developer access (which has restricted access). What is the result?

Answers:

A) Based on administrator access, the user does NOT have any limitations

B) The user is unable to perform any tasks in the instance until the contradiction is fixed

C) An error is produced because this is NOT a compatible configuration

D) The user is limited to the rights granted to the Developer group

Solution: A

Question: 4 *One Answer Is Right*

In which three execution status states can a campaign business practitioner start a workflow? (Choose three.)

Answers:

A) Finished

B) Being Edited

C) Paused

D) Error

Solution: A, B, C

Explanation:

Explanation: Reference:
https://docs.campaign.adobe.com/doc/AC/en/
WKF_General_operation_Executing_a_workflow.html

Question: 5 *One Answer Is Right*

For which reason does a fork activity produce an error?

Answers:

A) The fork has no inbound transition

B) The fork's outbound transition are out of sequence

C) The fork has an inbound transition that has zero results

D) The fork's outbound transition has no connected activity

Solution: D

Question: 6 *One Answer Is Right*

What three file formats are available when exporting data via a data extraction from adobe campaign? (Choose three.)

Answers:

A) PDF

B) CSV

C) XLS

D) XML

E) TXT

F) DOC

Solution: B, D, E

Explanation:

Explanation: Reference:
https://docs.adobe.com/content/help/en/campaign-standard/using/managing-processes-and- data/data-management-activities/extract-file.html

Question: 7 *One Answer Is Right*

How should a campaign business practitioner send a proof to a seed list?

Answers:

A) Proof's are sent automatically to seed lists before a delivery is sent

B) Select the appropriate target in the send proof menu

C) A proof cannot be sent to the seed list, only the actual delivery

D) Enter a seed list address in the "To:" text box

Solution: B

Explanation:

Explanation: Reference:
https://docs.campaign.adobe.com/doc/AC/en/
DLV_Using_seed_addresses_About_seed_addresses.html

Question: 8 *One Answer Is Right*

What should a campaign business practitioner add to a workflow that needs an additional data appended to target population?

Answers:

A) Add a list update to the workflow

B) Add a data import step to the workflow

C) Add an enrichment to the workflow

D) Add an intersection to the workflow

Solution: B

Explanation:

Explanation: Reference:
https://docs.campaign.adobe.com/doc/AC/en/
WKF_Repository_of_activities_Targeting_activities.html#Adding
_data

Question: 9 *One Answer Is Right*

A workflow supervisors group is assigned to a workflow. In
which instance is the workflow supervisors group notified?

Answers:

A) When a workflow deploys

B) When a workflow has thrown an error

C) When a workflow is taking a long time to run

D) When a delivery has changed

Solution: B

Explanation:

Explanation: Reference:
https://helpx.adobe.com/campaign/standard/automating/usin
g/executing-a-workflow.html

Question: 10 *One Answer Is Right*

Which is mandatory when creating a new campaign?

Answers:

A) Label

B) Start date

C) Plan

D) Channel

Solution: A

Explanation:

Explanation: Reference:
https://docs.campaign.adobe.com/doc/AC/en/
CMP_Orchestrate_campaigns_Setting_up_marketing_campaigns.
html

Chapter 13: AD0-E102 - Adobe Experience Manager BusinessPractitioner

Exam Guide

Adobe Experience Manager Business Practitioner AD0-E102 Exam:

Adobe Experience Manager Business Practitioner AD0-E102 Exam is related to Adobe Certified Expert Certification. This exam validates the Candidate knowledge and skills of AEM features and capabilities needed to engage developers to find business solutions, a deep insight into modules such as Assets, Sites and Forms. It also verifies the Candidate strong understanding of what modules are present and be able to recommend Adobe Experience Cloud solutions to meet business needs.

AD0-E102 Exam topics:

Candidates must know the exam topics before they start of preparation. Because it will really help them in hitting the core. Our **Adobe AD0-E102 dumps** will include the following topics:

- Education 16%
- Architecture 26%
- Business Analysis 48%
- Solution 10%

Certification Path:

There is no prerequisite for this Adobe AD0-E102 exam.

Who should take the AD0-E102 exam:

The Adobe Certified Expert certification is an internationally-recognized validation that identifies persons who earn it as possessing skilled in Adobe Experience Manager. If a candidate wants significant improvement in career growth needs enhanced knowledge, skills, and talents. The Adobe Certified Expert certification provides proof of this advanced knowledge and skill. If a candidate has knowledge of associated technologies and skills that are required to pass Adobe AD0-E102 Exam then he should take this exam.

How to study the AD0-E102 Exam:

There are two main types of resources for preparation of certification exams first there are the study guides and the books that are detailed and suitable for building knowledge from ground up then there are video tutorial and lectures that can somehow ease the pain of through study and are comparatively less boring for some candidates yet these demand time and concentration from the learner. Smart Candidates who want to build a solid foundation in all exam topics and related technologies usually combine video lectures with study guides to reap the benefits of both but there is one crucial preparation tool as often overlooked by most candidates the practice exams. Practice exams are built to make students comfortable with the real exam environment. Statistics have shown that most students fail not due to that preparation but due to exam anxiety the fear of the unknown. Certification-questions.com expert team recommends you to prepare some notes on these topics along with it don't forget to practice **Adobe AD0-E102 dumps** which been written by our expert team, Both these will help you a lot to clear this exam with good marks.

How much AD0-E102 Exam Cost:

The price of AD0-E102 exam is $180 USD.

How to book the AD0-E102 Exam:

These are following steps for registering the Adobe AD0-E102 exam.
Step 1: Visit to Pearson Exam Registration
Step 2: Signup/Login to Pearson VUE account
Step 3: Search for Adobe AD0-E102 Exam Certifications Exam
Step 4: Select Date, time and confirm with payment method

What is the duration, language, and format of the AD0-E102 Exam:

- Format: Multiple choices, multiple answers
- Length of Examination: 80 minutes
- Number of Questions: 50
- Passing score: 60%

The benefit in Obtaining the AD0-E102 Exam Certification:

- Resumes with Adobe Certified Expert certifications get noticed and fast-tracked by hiring managers.
- Adobe Certified Expert recognition and respect from colleagues and employers.
- Adobe Certified Expert receives exclusive updates on Adobe's latest products and innovations.
- Adobe Certified Expert can display Adobe certification logos on their business cards, resumes, and websites to leverage the power of the Adobe brand.

Difficulty in writing AD0-E102 Exam:

Candidates face many problems when they start preparing for the Adobe AD0-E102 exam. If a candidate wants to prepare his for the Adobe AD0-E102 exam without any problem and get good grades in the exam. Then they have to choose the best **Adobe AD0-E102 dumps** for real exam questions practice. There are many websites that are offering the latest Adobe AD0-E102 exam questions and answers but these questions are not verified by Adobe certified experts and that's why many are failed in their just first attempt. Certification-questions is the best platform which provides the candidate with the necessary Adobe AD0-E102 questions that will help him to pass the Adobe AD0-E102 exam on the first time. The candidate will not have to take the Adobe AD0-E102 exam twice because with the help of **Adobe AD0-E102 dumps** Candidate will have every valuable material required to pass the Adobe AD0-E102 exam. We are providing the latest and actual questions and that is the reason why this is the one that he needs to use and there are no chances to fail when a candidate will have valid braindumps from Certification-questions. We have the guarantee that the questions that we have will be the ones that will pass candidate in the Adobe AD0-E102 exam in the very first attempt.

For more info visit::

Adobe AD0-E102 Exam Reference

Sample Practice Test for AD0-E102

Question: 1 *One Answer Is Right*

A company has an existing English language site for the Canadian market. It is planning to create a new site for the US market. While most of the control of the current site can be reused for the new site, how would you create the new site in the most efficient manner?

Answers:

A) Create a live copy from the (ca/en) root to (us/en)

B) Duplicate the site root(ca/en), then move it to the new regional root

C) Copy the site root (ca/en) and paste it in the regional root (us/en)

D) Create a site (us/en) and define the redirect to (ca/en)

Solution: A

Question: 2 *One Answer Is Right*

An author would like to display an AI-summarized version of an article. Which method would you recommend to achieve this?

Answers:

A) Create a variation of a master content fragment

B) Modify a component to display a shortened form of the article

C) Create a variation of an experience fragment

D) Implement a workflow to generate a summarized version

Solution: A

Explanation:

Explanation: Reference: https://helpx.adobe.com/experience-manager/using/content-fragments.html

Question: 3 *One Answer Is Right*

A company plans to develop a set of pages with the same design and structure. The only difference between the pages is the content inside the body. What is the best approach to develop the pages?

Answers:

A) Create a specific page template for each page with associated components in each body

B) Create a page template for all pages with a layout container in the body

C) Create a page template and put the rich-text (RTE) in the body

D) Use the out-of-the-box Reference component to allow freedom in context editing

Solution: B

Question: 4 *One Answer Is Right*

Refer to the exhibit. Which is an AEM page mode?

Answers:

A) Timewarp

B) Publish

C) Deploy

D) Review

Solution: A

Explanation:

Explanation: Reference: https://helpx.adobe.com/experience-manager/6-3/sites/authoring/using/author-environment-tools.html#main-pars_title_20

Question: 5 *One Answer Is Right*

When a user requests a cacheable document from the AEM Dispatcher, what will the Dispatcher check to access whether the document exists in the web server file system? Choose two.

Answers:

A) If the document is not cached, the Dispatcher requests the document form the AEM instance

B) If the document is not cached, the Dispatcher returns a 404 error response

C) If the document is cached, the AEM Dispatcher returns the file from the CDN Cache

D) If the document is cached, the AEM Dispather requests the cached file after validation

Solution: A, C

Explanation:

Explanation: Reference:
https://docs.adobe.com/content/help/en/experience-manager-dispatcher/using/dispatcher.html

Question: 6 *One Answer Is Right*

A client wishes to moderate communication between their customers and their in-house product experts, expecting a very large amount of user-generated data. How can this be accomplished?

Answers:

A) Enable reverse replication

B) Create a forum and Q&A site using AEM Communities

C) Implement a contact us page using AEM Forms

D) Conduct an outreach marketing event with Adobe Campaign

Solution: B

Question: 7 *One Answer Is Right*

Your company is launching a website with customers from around the world. What capabilities of AEM Sites would allow you to make the content more relevant to the largest amount of customers? Choose two.

Answers:

A) Translation Framework

B) Layouting Mode

C) Editable Templates

D) Multi-Site Manager

E) Launches

Solution: B, D

Question: 8 *One Answer Is Right*

What is the recommended method to view a page as it was in a previous point in time?

Answers:

A) Revert pages to their previous versions

B) Use Timewarp to simulate the published state of a page at specific times in the past

C) Create a workflow to take screen captures of pages and store them in Assets

D) Keep nightly backups of CRX and restore as needed

Solution: B

Explanation:

Explanation: Reference: https://helpx.adobe.com/experience-manager/6-3/sites/authoring/using/working-with-page-versions.html

Question: 9 *One Answer Is Right*

You want to create a grouping of all the assets under a specific category that are contained in various location across the DAM. What is the best way to achieve this?

Answers:

A) Use the Dropzone feature

B) Use static references to assets

C) Use Smart collections

D) Use static references to folders

Solution: C

Explanation:

Explanation: Reference: https://helpx.adobe.com/experience-manager/6-3/assets/using/managing-collections-touch-ui.html

Question: 10 *One Answer Is Right*

A client is building out their website using several different content pages web specific components to be used on each page. What is the recommended system to restrict components on specific templates?

Answers:

A) Set ACL permissions

B) Define components in page properties

C) Use content policies

D) Use the Responsive Grid Edit dialog

Solution: C

Explanation:

Explanation: Reference:
https://forums.adobe.com/thread/2451356

Chapter 14: AD0-E103 - Adobe Experience Manager Developer

Exam Guide

Adobe Experience Manager Developer AD0-E103 Exam:

Adobe Experience Manager Developer AD0-E103 Exam is related to Adobe Certified Expert Certification. This exam validates the ability to Create, extend and configure templates and components, build and configure OSGi components and services, use Sling and JCR API. It also verifies a Candidate have general knowledge about building using Apache Maven and be able to set-up their own environment (e.g., Java SDK, and AEM), Test applications and troubleshoot AEM projects.

AD0-E103 Exam topics:

Candidates must know the exam topics before they start of preparation. Because it will really help them in hitting the core. Our **Adobe AD0-E103 dumps** will include the following topics:

- Installation and Configuration of AEM 10%
- Templates and Components 40%
- OSGi Services 20%
- Packaging and Deploying AEM projects 10%
- Troubleshooting AEM projects 20%

Certification Path:

There is no prerequisite for this Adobe AD0-E103 exam.

Who should take the AD0-E103 exam:

The Adobe Certified Expert certification is an internationally-recognized validation that identifies persons who earn it as possessing skilled in Adobe Experience Manager. If a candidate wants significant improvement in career growth needs enhanced knowledge, skills, and talents. The Adobe Certified Expert certification provides proof of this advanced knowledge and skill. If a candidate has knowledge of associated technologies and skills that are required to pass Adobe AD0-E103 Exam then he should take this exam.

How to study the AD0-E103 Exam:

There are two main types of resources for preparation of certification exams first there are the study guides and the books that are detailed and suitable for building knowledge from ground up then there are video tutorial and lectures that can somehow ease the pain of through study and are comparatively less boring for some candidates yet these demand time and concentration from the learner. Smart Candidates who want to build a solid foundation in all exam topics and related technologies usually combine video lectures with study guides to reap the benefits of both but there is one crucial preparation tool as often overlooked by most candidates the practice exams. Practice exams are built to make students comfortable with the real exam environment. Statistics have shown that most students fail not due to that preparation but due to exam anxiety the fear of the unknown. Certification-questions.com expert team recommends you to prepare some notes on these topics along with it don't forget to practice **Adobe AD0-E103 dumps** which been written by our expert team, Both these will help you a lot to clear this exam with good marks.

How much AD0-E103 Exam Cost:

The price of AD0-E103 exam is $180 USD.

How to book the AD0-E103 Exam:

These are following steps for registering the Adobe AD0-E103 exam.

Step 1: Visit to Pearson Exam Registration

Step 2: Signup/Login to Pearson VUE account

Step 3: Search for Adobe AD0-E103 Exam Certifications Exam

Step 4: Select Date, time and confirm with payment method

What is the duration, language, and format of the AD0-E103 Exam:

- Format: Multiple choices, multiple answers
- Length of Examination: 90 minutes
- Number of Questions: 50
- Passing score: 60%

The benefit in Obtaining the AD0-E103 Exam Certification:

- Resumes with Adobe Certified Expert certifications get noticed and fast-tracked by hiring managers.
- Adobe Certified Expert recognition and respect from colleagues and employers.
- Adobe Certified Expert receives exclusive updates on Adobe's latest products and innovations.
- Adobe Certified Expert can display Adobe certification logos on their business cards, resumes, and websites to leverage the power of the Adobe brand.

Difficulty in writing AD0-E103 Exam:

This exam is very difficult especially for those who have not on the job experience as an Adobe Certified Expert. Candidates can not pass this exam with only taking courses because courses do not provide the knowledge and skills that are necessary to pass this exam. Certification-questions.com is the best platform for those who want to pass Adobe 9A0-381 with good grades in no time. Certification-questions.com provides the latest **Adobe 9A0-381 dumps** that will immensely help candidates to get good grades in their final Adobe 9A0-381 exam. Certification-questions.com is one of the best study sources to provide the most updated **Adobe 9A0-381 Dumps** with our Actual Adobe 9A0-381 Exam Questions PDF. Candidate can rest guaranteed that they will pass their Adobe 9A0-381 Exam on the first attempt. We will also save candidates valuable time. Certification-questions Dumps help to pass the exam easily. Candidates can get all real questions from Certification-questions. One of the best parts is we also provide most updated Adobe Certified Expert Exam study materials and we also want a candidate to be able to access study materials easily whenever they want. So, We provide all our Adobe 9A0-381 exam questions in a very common PDF format that is accessible from all devices.

For more info visit::

Adobe AD0-E103 Exam Reference

Sample Practice Test for AD0-E103

Question: 1 *One Answer Is Right*

A developer needs to create a banner component. This component shows an image across the full width of the page. A title is shown on top of the image. This text can be aligned to the left, middle, or right. The core components feature a teaser component which matches almost all requirements, but not all. What is the most maintainable way for the developer to implement these requirements?

Answers:

A) Use and configure the teaser core component.

B) Create a new custom component from scratch.

C) Overlay the teaser core component.

D) Inherit from the teaser core component.

Solution: A

Explanation:

Explanation: Reference:
https://helpx.adobe.com/es/experience-manager/kt/sites/using/getting-started-wknd-tutorial-develop/part7.html

Question: 2 *One Answer Is Right*

A developer is creating a custom component that shows a list of pages. For each page, the following items must be shown: - Title of the page - Description of the page - A button with fixed text "Read more" that must be translatable All of the above fields must be wrapped in a

tag. The logic for obtaining the list of pages must be reusable for future components. Which snippet should the developer use to meet these requirements?

Answers:

A)

```
<sly data-sly-use.model="com.example.Component"/>
<div data-sly-list="${model.pages}">
    <p>${item.title}</p>
    <p>${item.description}</p>
    <a href= "${item.link}">${"Read more" @ .translate} <a>
</div>
```

B)

```
<sly data-sly-load.model="com.example.Component"/>
<div data-sly-list.page="${model.pages}">
    <p>${page.title}</p>
    <p>${page.description}</p>
    <a href= "${page.link}">${"Read more" @ i18n} <a>
</div>
```

C)

```
<sly data-sly-use.model="com.example.Component"/>
<div data-sly-list="${model.pages}">
    <p>${item.title}</p>
    <p>${item.description}</p>
    <a href= "${item.link}">${"Read more" @ .i18n} <a>
</div>
```

D)

```
<sly data-sly-use.model="com.example.Component"/>
<div data-sly-list="${model.pages}">
    <p>${model.title}</p>
    <p>${model.description}</p>
    <a href= "${item.link}">${"Read more" @ .translate} <a>
</div>
```

Solution: B

Question: 3 *One Answer Is Right*

A developer is working on a complex project with multiple bundles. One bundle provides an OSGi service for other bundles. Which two options are necessary to ensure that the other bundles can reference that OSGi service? (Choose two.)

Answers:

A) The bundles consuming the service need to import the fully qualified name of the service interface.

B) The service needs to correctly declare metatype information.

C) The bundle providing the service needs to contain a whitelist of allowed consumer bundles.

D) The bundle providing the service needs to contain an adequate SCR descriptor file.

E) The bundle providing the service needs to export the java package of the service interface.

Solution: C, E

Question: 4 *One Answer Is Right*

The structure section of an editable template has a locked component. What happens to the content of that component when a developer unlocks it?

Answers:

A) The content stays in the same place but it ignored on pages using the template.

B) The content is moved to the initial section of the editable template.

C) The content is deleted after confirmation from the template author.

D) The content is copied to the initial section of the editable template.

Solution: B

Explanation:

Explanation: Reference: https://helpx.adobe.com/experience-manager/6-3/sites/developing/using/page-templates-editable.html

Question: 5 *One Answer Is Right*

Which log file contains AEM application request and response entries?

Answers:

A) response.log

B) request.log

C) history.log

D) audit.log

Solution: B

Explanation:

Explanation: Reference: http://www.sgaemsolutions.com/2017/04/aem-logs-in-detail-part-1.html

Question: 6 *One Answer Is Right*

A developer identifies that some requests for the page /content/sampleproject/page.html take longer that other requests for the same page. Refer to the $DOCROOT/content/sampleproject directory below.

```
[user@group /opt/dispatcher/cache/content/sampleproject ]$ ls -la
total 2
drwxr-xr-x. 5 apache apache 4096 Feb 11 11:41 .
drwxr-xr-x. 3 apache apache 4096 Nov 29 16:07 ..
drwxr-xr-x. 4 apache apache 4096 Feb 7 03:21 page.html
-rw-r--r--. 1 apache apache 0 Feb 7 03:19 .stat
```

The dispatcher.log file contains the following lines:

```
[Wed Feb 13 13:14:04 2012]  [D]  [1376(1532)]  checking  [/libs/cq/security/userinfo/json]
[Wed Feb 13 13:14:04 2012]  [D]  [1376(1532)]  Caching disabled due to query string: tracking_id=1350373444666
[Wed Feb 13 13:14:04 2012]  [D]  [1376(1532)]  cache-action for [/libs/cq/security/userinfo/json]: NONE
```

How should the developer make sure that the page is always cached?

Answers:

A) Modify the dispatcher.any file to contain the following lines:

```
/filter
   {
    ...
    /0023 { /type "allow" /url "/content/*/*/html" /params "tracking_id"}
    ...
   }
```

B) Modify the dispatcher.any file to contain the following lines:

```
/rules
   {
    ...
    /0000 { /glob "*" /type "allow" /params "tracking_id"}
    ...
   }
```

C) Modify the dispatcher.any file to contain the following lines:

```
/filter
   {
    ...
    /0023 { /type "allow" /url "/content/*/*.html?tracking_id=*"}
    ...
   }
```

D) Modify the dispatcher.any file to contain the following lines:

```
/ignoreUrlParams
    {
    ...
    /0002 { /glob "tracking_id"/type "allow" }
    ...
    }
```

Solution: C

Question: 7 *One Answer Is Right*

A developer creates a Sling Servlet. The Sling Servlet is bound to a path (/service/sling/sample). Refer to the resulting code below.

```
@Component (immediate=true,service = {Servlet.class})
@SlingServletPaths(value = { "/service/sling/sample" })
```

What should the developer do to make the servlet access controlled using the default ACLs?

Answers:

A) Use @SlingServletResourceTypes instead of @SlingServletPaths.

B) Modify @SlingServletPaths(value = {"/bin/sling/sample" }).

C) Add @SlingServletName(servletName = "AccessControlServlet") annotation.

D) Add @SlingServletPrefix(value = "/apps") annotation.

Solution: A

Explanation:

Explanation: Reference:
https://sling.apache.org/documentation/the-sling-engine/servlets.html#caveats-when-binding- servlets-by-path

Question: 8 *One Answer Is Right*

A developer wants to extend AEM Core Components to create a custom Carousel Component. How should the developer extend the Core Components?

Answers:

A) Make changes to the original component and assign a component group.

B) Use the sling:resourceSuperType property to point to the core component.

C) Use the sling:resourceType property to point to the core component.

D) Copy the Core Carousel component to /apps/ folder.

Solution: D

Question: 9 *One Answer Is Right*

A developer wants to change the log level for a custom API. Which OSGi configuration should the developer modify?

Answers:

A) Apache Sling Logging Configuration

B) Apache Sling Log Tracker Service

C) Apache Sling Logging Writer Configuration

D) Adobe Granite Log Analysis Service

Solution: A

Explanation:

Explanation: Reference:
https://docs.adobe.com/content/help/en/experience-manager-64/deploying/configuring/osgi- configuration-settings.html

Question: 10 *One Answer Is Right*

Refer to the following four Client Library Folders.

```
html
One
- categories="[library.one]"
- dependencies="[library.three, library.four]"
- embed="[library.two]"


Two
- categories="[library.two]"


Three
- categories="[library.three]"
- dependencies="[library.four]"


Four
- categories="[library.four]"
```

A developer uses the following:

```
html
<data-sly-call="${clientlib.css @ categories='library.one'}"/>
```

What is the resulting HTML?

Answers:

A)

```
<link rel="stylesheet" href="library.one.css">
<link rel="stylesheet" href="library.three.css">
<link rel="stylesheet" href="library.four.css">
```

B)

```
<link rel="stylesheet" href="library.two.css">
<link rel="stylesheet" href="library.one.css">
<link rel="stylesheet" href="library.three.css">
```

C)

```
<link rel="stylesheet" href="library.four.css">
<link rel="stylesheet" href="library.three.css">
<link rel="stylesheet" href="library.one.css">
```

D)

```
<link rel="stylesheet" href="library.three.css">
<link rel="stylesheet" href="library.four.css">
<link rel="stylesheet" href="library.one.css">
```

Solution: A

Chapter 15: AD0-E104 - Adobe Experience Manager Architect

Exam Guide

Adobe Experience Manager Architect AD0-E104 Exam:

Adobe Experience Manager Architect AD0-E104 Exam is related to Adobe Certified Expert Certification. This exam validates the Candidate knowledge and skills of web and content-centric application development. It also deals with the ability to design, configure and deploy an AEM application in multiple modes (author, publisher) along with optimal dispatcher configurations, analyze business requirements and follow best practices to produce an executable solution, advise and participate in all phases of an AEM project.

AD0-E104 Exam topics:

Candidates must know the exam topics before they start of preparation. Because it will really help them in hitting the core. Our **Adobe AD0-E104 dumps** will include the following topics:

- Business Requirements 25%
- Architecture and Design 40%
- Configuration and Deployment 22%
- System Maintenance 13%

  Certification Path:

There is no prerequisite for this Adobe AD0-E104 exam.

Who should take the AD0-E104 exam:

The Adobe Certified Expert certification is an internationally-recognized validation that identifies persons who earn it as possessing skilled in Adobe Experience Manager. If a candidate wants significant improvement in career growth needs enhanced knowledge, skills, and talents. The Adobe Certified Expert certification provides proof of this advanced knowledge and skill. If a candidate has knowledge of associated technologies and skills that are required to pass Adobe AD0-E104 Exam then he should take this exam.

How to study the AD0-E104 Exam:

There are two main types of resources for preparation of certification exams first there are the study guides and the books that are detailed and suitable for building knowledge from ground up then there are video tutorial and lectures that can somehow ease the pain of through study and are comparatively less boring for some candidates yet these demand time and concentration from the learner. Smart Candidates who want to build a solid foundation in all exam topics and related technologies usually combine video lectures with study guides to reap the benefits of both but there is one crucial preparation tool as often overlooked by most candidates the practice exams. Practice exams are built to make students comfortable with the real exam environment. Statistics have shown that most students fail not due to that preparation but due to exam anxiety the fear of the unknown. Certification-questions.com expert team recommends you to prepare some notes on these topics along with it don't forget to practice **Adobe AD0-E104 dumps** which been written by our expert team, Both these will help you a lot to clear this exam with good marks.

How much AD0-E104 Exam Cost:

The price of AD0-E104 exam is $180 USD.

How to book the AD0-E104 Exam:

These are following steps for registering the Adobe AD0-E104 exam.

Step 1: Visit to Pearson Exam Registration

Step 2: Signup/Login to Pearson VUE account

Step 3: Search for Adobe AD0-E104 Exam Certifications Exam

Step 4: Select Date, time and confirm with payment method

What is the duration, language, and format of the AD0-E104 Exam:

- Format: Multiple choices, multiple answers
- Length of Examination: 120 minutes
- Number of Questions: 60
- Passing score: 60%

The benefit in Obtaining the AD0-E104 Exam Certification:

- Resumes with Adobe Certified Expert certifications get noticed and fast-tracked by hiring managers.
- Adobe Certified Expert recognition and respect from colleagues and employers.
- Adobe Certified Expert receives exclusive updates on Adobe's latest products and innovations.
- Adobe Certified Expert can display Adobe certification logos on their business cards, resumes, and websites to leverage the power of the Adobe brand.

Difficulty in writing AD0-E104 Exam:

Mostly job holder candidates give a short time to their study and want to pass the exam with good marks. Thereby we have many ways to prepare and practice for exams in a very short time that help the candidates to ready for exams in a very short time without any tension. Candidates can easily prepare Adobe 9A0-395 exams from Certification-questions because we are providing the best Adobe 9A0-395 dumps which are verified by our experts. Certification-questions has always verified and updated **Adobe 9A0-395 dumps** that helps the candidate to prepare his exam with little effort in a very short time. We also provide latest and relevant study guide material which is very useful for a candidate to prepare easily for **Adobe 9A0-395 dumps**. Candidate can download and read the latest dumps in PDF and VCE format. Certification-questions is providing real questions of **Adobe 9A0-395 dumps**. We are very fully aware of the importance of student time and money that's why Certification-questions give the candidate the most astounding brain dumps having all the inquiries answer outlined and verified by our experts.

For more info visit::

Adobe AD0-E104 Exam Reference

Sample Practice Test for AD0-E104

Question: 1 *One Answer Is Right*

Which solution is required to use segments from Adobe Analytics in AEM in authoring mode?

Answers:

A) Adobe Launch

B) Adobe Target

C) Adobe Experience Cloud ID Service

D) Adobe Audience Manager

Solution: B

Question: 2 *One Answer Is Right*

A content author creates an article page in AEM. A notification must be sent to a third-party system. An Architect needs to make sure there are multiple attempts to deliver the notification in the event the third-party system is temporarily unavailable. How should the Architect meet this requirement?

Answers:

A) Sling Pipes

B) Sling Jobs

C) Event Handlers

D) Event Listeners

Solution: C

Question: 3 *One Answer Is Right*

When should Closed User Groups (CUGs) be used?

Answers:

A) To create an immutable user group

B) To limit access to specific pages on a published site

C) To create a user group with custom permissions

D) To limit users in order to improve performance of a published site

Solution: B

Question: 4 *One Answer Is Right*

A customer has a large MSM setup with more than 30 languages and 150 websites: The content authors are frequently updating the master content. MSM is configured with the "Push on modify" rollout configuration so that any changes are immediately pushed to all live copies. The customer is complaining that they encounter performance issues on the AEM Author. What should the Architect do?

Answers:

A) Configure MSM to use "Standard rollout config" with the "On Modification" trigger

B) Configure MSM to use "Push on modify (shallow)"

C) Configure MSM to use "Standard rollout config"

D) Remove the referencesUpdate action from the "Push on modify"

Solution: C

Question: 5 *One Answer Is Right*

Which AEM pre-deployment action should the Architect do after final UAT testing but prior to a live production launch?

Answers:

A) Forecast overall visitor traffic

B) Stress test all Dispatchers and Publish instances

C) Execute relevant AEM licensing and maintenance agreements

D) Understand migration requirements

Solution: B

Question: 6 *One Answer Is Right*

Which node store should be configured to store a large number of binaries?

Answers:

A) Default JCR Repository

B) Segment node store

C) File Data store

D) Document node store

Solution: C

Question: 7 *One Answer Is Right*

A client's marketing pages are generally slow to load which is causing a significant drop in sales. All other AEM pages load within expected performance guidelines regardless of whether the visitor is being served the desktop or mobile experiences. The marketing pages typically get slower when multiple external campaigns such as Facebook or AdWords drive traffic to those pages. The page performance tends to dip during high traffic periods Internal campaign clicks such as those from hero images use similar campaign codes as external campaign traffic. What should the Architect do to resolve this issue?

Answers:

A) Convert the marketing pages to a responsive design instead of an adaptive design

B) Set AEM to use GZIP compression instead of the web servers compression

C) Add a new dispatcher farm to assist with the clients marketing pages

D) Modify the dispatcher.any files section to ignore campaign-based URL parameters

Solution: B

Question: 8 *One Answer Is Right*

A customer's photo gallery site uses query parameters to filter photo search results. The site experiences high AEM Publish server load when users filter photos on the site. The customer would like to identify the cause of this issue. What should the Architect investigate first?

Answers:

A) Dispatcher load balancing configuration

B) Volume of assets being loaded at a time

C) Cache-Control Headers in dispatcher.any

D) ignoreUrlParams configuration in dispatcher.any

Solution: A

Question: 9 *One Answer Is Right*

A customer is using an AEM 6.4 core image component for their website. They have an additional requirement to render the title of the associated tags with the image. How should the Architect implement the new requirement?

Answers:

A) Using the Java-use class

B) Using annotations

C) Using a Sling Model

D) Using HTL

Solution: C

Question: 10 *One Answer Is Right*

Why must Service Users be created?

Answers:

A) To allow mapping OSGi services to users

B) To allow administration access to users

C) To restrict mapping OSGi services to groups

D) To restrict mapping OSGi services to admins

Solution: D

Chapter 16: AD0-E105 - Adobe Experience Manager Lead Developer

Exam Guide

Adobe Experience Manager Lead Developer AD0-E105 Exam:

Adobe Experience Manager Lead Developer AD0-E105 Exam is related to Adobe Certified Expert Certification. This exam validates the Candidate ability to configure, dispatcher (caching, security, with CDN), implement a content structure and taxonomy for a project, employ security guidelines, configure AEM as a headless CMS (e.g. using the SPA editor, creating SPA components), create asset schemas, maintain build processes, translate user requirements into technical designs and implement requirements for automated testing.

AD0-E105 Exam topics:

Candidates must know the exam topics before they start of preparation. Because it will really help them in hitting the core. Our **Adobe AD0-E105 dumps** will include the following topics:

- Installation, Integration, and Configuration of AEM 16%
- Configuring, managing, and deploying AEM projects 22%
- Workflows, Templates, and Components 20%
- OSGi Components, Services, and Configuration 20%

- Maintaining AEM projects 22%

  Certification Path:

There is no prerequisite for this Adobe AD0-E105 exam.

Who should take the AD0-E105 exam:

The Adobe Certified Expert certification is an internationally-recognized validation that identifies persons who earn it as possessing skilled in Adobe Experience Manager. If a candidate wants significant improvement in career growth needs enhanced knowledge, skills, and talents. The Adobe Certified Expert certification provides proof of this advanced knowledge and skill. If a candidate has knowledge of associated technologies and skills that are required to pass Adobe AD0-E105 Exam then he should take this exam.

How to study the AD0-E105 Exam:

There are two main types of resources for preparation of certification exams first there are the study guides and the books that are detailed and suitable for building knowledge from ground up then there are video tutorial and lectures that can somehow ease the pain of through study and are comparatively less boring for some candidates yet these demand time and concentration from the learner. Smart Candidates who want to build a solid foundation in all exam topics and related technologies usually combine video lectures with study guides to reap the benefits of both but there is one crucial preparation tool as often overlooked by most candidates the practice exams. Practice exams are built to make students comfortable with the real exam environment. Statistics have shown that most students fail not due to that preparation but due to exam anxiety the fear of the unknown. Certification-questions.com expert team recommends you to prepare some notes on these topics along with it don't forget to practice

Adobe AD0-E105 dumps which have been written by our expert team, Both these will help you a lot to clear this exam with good marks.

How much AD0-E105 Exam Cost:

The price of AD0-E105 exam is $180 USD.

How to book the AD0-E105 Exam:

These are following steps for registering the Adobe AD0-E105 exam.
Step 1: Visit to Pearson Exam Registration
Step 2: Signup/Login to Pearson VUE account
Step 3: Search for Adobe AD0-E105 Exam Certifications Exam
Step 4: Select Date, time and confirm with payment method

What is the duration, language, and format of the AD0-E105 Exam:

- Format: Multiple choices, multiple answers
- Length of Examination: 100 minutes
- Number of Questions: 50
- Passing score: 60%

The benefit in Obtaining the AD0-E105 Exam Certification:

- Resumes with Adobe Certified Expert certifications get noticed and fast-tracked by hiring managers.
- Adobe Certified Expert recognition and respect from colleagues and employers.
- Adobe Certified Expert receives exclusive updates on Adobe's latest products and innovations.
- Adobe Certified Expert can display Adobe certification logos on their business cards, resumes, and websites to leverage the power of the Adobe brand.

Difficulty in writing AD0-E105 Exam:

This Adobe AD0-E105 exam is very difficult to prepare. Because it requires all candidate attention with practice. So, if Candidate wants to pass this Adobe AD0-E105 exam with good grades then he has to choose the right preparation material. By passing the Adobe AD0-E105 exam can make a lot of difference in your career. Many Candidates wants to achieve success in the Adobe AD0-E105 exam but they are failing in it. Because of their wrong selection but if the candidate can get valid and latest Adobe AD0-E105 study material then he can easily get good grades in the Adobe AD0-E105 exam. Certification-questions providing many Adobe AD0-E105 exam questions that help the candidate to get success in the Adobe AD0-E105 test. Our **Adobe AD0-E105 dumps** specially designed for those who want to get their desired results in the just first attempt. **Adobe AD0-E105 dump** questions provided by Certification-questions make candidate preparation material more impactful and the best part is that the training material provided by Certification-questions for Adobe AD0-E105 exams are designed by our experts in the several fields of the IT industry.

For more info visit::

Adobe AD0-E105 Exam Reference

Chapter 17: AD0-E106 - Adobe Experience Manager Dev/Ops Engineer

Exam Guide

Adobe Experience Manager Dev/Ops Engineer AD0-E106 Exam:

Adobe Experience Manager Dev/Ops Engineer AD0-E106 Exam is related to Adobe Certified Expert Certification. This exam validates the ability to install, configure, monitor, operate, and troubleshoot AEM (V6.3 and higher). It also deals with the ability to configure continuous integration/continuous deployment tasks.

AD0-E106 Exam topics:

Candidates must know the exam topics before they start of preparation. Because it will really help them in hitting the core. Our **Adobe AD0-E106 dumps** will include the following topics:

- Install and Configure AEM 24%
- Troubleshoot AEM Environments 22%
- Install and configure Dispatcher 13%
- Build and Deployments 15%
- Maintenance and Operations 26%

Certification Path:

There is no prerequisite for this Adobe AD0-E106 exam.

Who should take the AD0-E106 exam:

The Adobe Certified Expert certification is an internationally-recognized validation that identifies persons who earn it as possessing skilled in Adobe Experience Manager. If a candidate wants significant improvement in career growth needs enhanced knowledge, skills, and talents. The Adobe Certified Expert certification provides proof of this advanced knowledge and skill. If a candidate has knowledge of associated technologies and skills that are required to pass Adobe AD0-E106 Exam then he should take this exam.

How to study the AD0-E106 Exam:

There are two main types of resources for preparation of certification exams first there are the study guides and the books that are detailed and suitable for building knowledge from ground up then there are video tutorial and lectures that can somehow ease the pain of through study and are comparatively less boring for some candidates yet these demand time and concentration from the learner. Smart Candidates who want to build a solid foundation in all exam topics and related technologies usually combine video lectures with study guides to reap the benefits of both but there is one crucial preparation tool as often overlooked by most candidates the practice exams. Practice exams are built to make students comfortable with the real exam environment. Statistics have shown that most students fail not due to that preparation but due to exam anxiety the fear of the unknown. Certification-questions.com expert team recommends you to prepare some notes on these topics along with it don't forget to practice **Adobe AD0-E106 dumps** which have been written by our expert team, Both these will help you a lot to clear this exam with good marks.

How much AD0-E106 Exam Cost:

The price of AD0-E106 exam is $180 USD.

How to book the AD0-E106 Exam:

These are following steps for registering the Adobe AD0-E106 exam.
Step 1: Visit to Pearson Exam Registration
Step 2: Signup/Login to Pearson VUE account
Step 3: Search for Adobe AD0-E106 Exam Certifications Exam
Step 4: Select Date, time and confirm with payment method

What is the duration, language, and format of the AD0-E106 Exam:

- Format: Multiple choices, multiple answers
- Length of Examination: 108 minutes
- Number of Questions: 54
- Passing score: 60%

The benefit in Obtaining the AD0-E106 Exam Certification:

- Resumes with Adobe Certified Expert certifications get noticed and fast-tracked by hiring managers.
- Adobe Certified Expert recognition and respect from colleagues and employers.
- Adobe Certified Expert receives exclusive updates on Adobe's latest products and innovations.
- Adobe Certified Expert can display Adobe certification logos on their business cards, resumes, and websites to leverage the power of the Adobe brand.

Difficulty in writing AD0-E106 Exam:

Adobe AD0-E106 Exam certification exam has a higher rank in the IT sector. Candidate can add a most powerful Adobe Certified Expert certification on their resume by passing Adobe AD0-E106 exam. AD0-E106 is a very challenging exam Candidate will have to work hard to pass this exam. With the help of Certification-questions provided the right focus and preparation material passing this exam is an achievable goal. Certification-Questions provide the most relevant and updated **Adobe AD0-E106 dumps**. Furthermore, We also provide the AD0-E106 practice test that will be much beneficial in the preparation. Our aims to provide the best **Adobe AD0-E106 pdf dumps**. We are providing all useful preparation materials such as **Adobe AD0-E106 dumps** that had been verified by the Adobe experts, **AD0-E106 dumps** and customer care service in case of any problem. These are things are very helpful in passing the exam with good grades.

For more info visit::

Adobe AD0-E106 Exam Reference

Sample Practice Test for AD0-E106

Question: 1 *One Answer Is Right*

How can the DevOps engineer prevent any user from logging in with the default admin credentials during startup of the AEM instance?

Answers:

A) Disable the OSGi web console login bundle using the production ready runmode

B) Change the default AEM admin password on initial setup

C) Configure the dispatcher to prevent access to /system/console

D) Update the default password in the OSGI configuration for the ISGI web console

Solution: C

Question: 2 *One Answer Is Right*

The DevOps Engineer sees many occurrences of org.apache.s;ing.api resource.loginException in the logs. Apache Sling Service User mapper Service is configured with a default user. Service user mapping is configured for the OSGi bundle, causing the exception. What is the root cause of these exceptions?

Answers:

A) Administrator resource resolvers have been disabled by configuration and the bundle deployed still relies on it.

B) The exception relate to failed login attempts with incorrect credentials.

C) The configured service user in not a system user or does not exist.

D) There is no service user mapping configured for the Java class causing the exception.

Solution: C

Question: 3 *One Answer Is Right*

A DevOps Engineer configures a delay in the out of the box online backup. What is he result of a delay that is too large?

Answers:

A) Too many file writes occur.

B) The backup takes more than 24 hours.

C) Excessive reads of the repository occur

D) The CPU usage is reduced too much

Solution: A

Question: 4 *One Answer Is Right*

A DevOps Engineer is deploying an AEM environment in a private network. How can the DevOps Engineer restrict other clients or applications from flushing the cache?

Answers:

A) The allowed section in the dispatcher configuration should only allow the publish instances private IPs

B) Set up client certificate authentication In the dispatcher configuration

C) Set permission on the publish instance configurations Control Lists to allow the publish instance private IPs.

D) Configure the .filter section of the dispatcher configuration with an allow rule to make sure the correct client's private IPs are allowed to flush the cache

Solution: B

Question: 5 *One Answer Is Right*

A business needs to remove a publish a instance due to contractual downsizing. Which action will prevent a rapid increase of errors in the author instance?

Answers:

A) Configure the dispatcher mapped to the publish instance being removed to display a maintenance page

B) Arrange a content freeze preventing access to the author instance while the publish instance is being removed.

C) Remove the dispatcher associated with the publish instance being removed

D) Delete the replication agent on the author instance mapped to the publish instance being removed

Solution: C

Question: 6 *One Answer Is Right*

On which instance should a flush agent be configured to prevent invalidation timing issues after invalidation?

Answers:

A) Load balancer

B) Publisher

C) Dispatcher

D) Author

Solution: A

Question: 7 *One Answer Is Right*

What supported when server should a DevOps Engineer use when setting up the dispatcher version 4.3.1 in a Unix environment.

Answers:

A) Nginx 1.14

B) Apache 2.4

C) Apache 2.0

D) IIS 7.5

Solution: A

Question: 8 *One Answer Is Right*

In what two ways can a DevOps engineer install a content package? (Choose two.)

Answers:

A) Use CRX Package Manager

B) Store the content package in the crx-quick/install folder in the filesystem

C) Store the content package in the crs-quickstart/app folder in the filesystem

D) Upload the package to content dam and start the installPaclagWorkflow

E) Upload the package through OSGi console

Solution: B, D

Question: 9 *One Answer Is Right*

When configuration agents, under which path of the repository are agent stored for the AEM author instance?

Answers:

A) /etc/replication/agenta.author

B) /etc/agent/replication/author

C) /etc/agent/replication.authot

D) /etc/author/agents/replication

Solution: D

Question: 10 *One Answer Is Right*

New content is not visible on the website when accessing it via the dispatcher. * Replication from author to publish works fine * Dispatcher flush agent is present under/replication/agents. Author on the Publish instance and enabled * The checkbox for Dispatcher flush agent confusion is ticked for enabled, when reviewed on the author instance Rules in the dispatcher flush agent is causing this issue?

Answers:

A) It is configured properly but located in the wrong path.

B) It does not have enough permission to receive the activation.

C) It is not configured properly in the dispatcher configuration.

D) It is configured properly but uses the incorrect transport user.

Solution: D

Chapter 18: AD0-E200 - Adobe Analytics Architect

Exam Guide

How to Prepare For Adobe Analytics Architect ADO-E200 Exam

Preparation Guide for Adobe Analytics Architect ADO-E200 Exam

Introduction

Adobe certification is the swift, economical, and official way to demonstrate your skills using a range of industry-standard software packages, so it's probably a good investment. You don't need to have training in these tools before taking the exam, but it is extremely advisable to have proper experience of the required Adobe Tools. Adobe's certification exams are developed following industry-accepted standards to ensure validity and reliability. We work with industry experts to create our exams, which represent real-world requirements and objectives for the job roles we certify.

Adobe Analytics Architect ADO-E200 Exam:

Adobe Analytics Architect **ADO-E200 Exam** is related to Adobe Certified master level Certification. This exam validates the Candidate knowledge and skills of adobe analytics Architect as per the business needs. It helps to demonstrate solution Design to manage report suites settings, data extraction, and relationships.

ADO-E200 Exam topics:

Candidates must know the exam topics before they start of preparation. Our Adobe **ADO-E200 dumps** will include the following topics:

1: Discover (18%)

- Demonstrate ability to audit sites
- Investigate client needs to build business requirements
- Given a scenario, create measurement framework

2: Solution Design (54%)

- Translate business requirements into variables and metrics
- Demonstrate ability to manage report suites settings
- Given a scenario, recommend the best methods for collecting data
- Given a scenario, determine the technical specifications to implement site tracking
- Create experience cloud users/groups

3: Post Implementation (28%)

- Given a set of requirements, validate tracking through browser developer tools and Adobe reports
- Apply procedural concepts to manage data extraction and relationships
- Demonstrate an understanding of a data governance model
- Given a scenario, manage data sources and connectors

- Given a set of requirements, validate tracking through browser developer tools and Adobe reports

Certification Path:

There is no prerequisite for this Adobe ADO-E200 exam. This exam is already expired, now you should appear for AD0-E207 Adobe Analytics Architect Master.

Who should take the ADO-E200 exam:

AD0-E200 exam been designed for the following target audience:

- Solution Architects
- Technical Manager
- Data Architect
- Analytics Engineer
- Analytics Strategist
- Multi Solutions Engineer

Those who plan to take ADO-E200 Exam should have minimum skillset as follows:-

- At a minimum, the candidate seeking to become certified as an Analytics Architect has a minimum of 3-4 years' experience analyzing and interpreting data as well as a basic understanding of JavaScript code and at least 3 years of Adobe Analytics experience. The Analytics Architect is responsible for translating business requirements into technical specifications, designing data architecture to map requirements to eVars, props, and events, and understands the interdependency between e-commerce systems, platforms, and the digital marketing life cycle.

Familiarity with the following technologies and environments:

- Adobe Analytics
- Analytics Admin Console
- Microsoft Excel
- Analytics Debuggers
- Adobe Target
- Adobe Campaign
- Tag Management Systems
- Browser Developer Tools
- HTML
- JavaScript
- jQuery and other JavaScript libraries
- On-site meetings with clients
- Any Web Analytics Application
- Adobe Experience Cloud

How to study the ADO-E200 Exam:

For Adobe Exam it is really important for any participant that they should have a proper study plan in place because simply reading without proper planning will not be adding any values. There are many sources are the there on the internet, mainly two types of resource can be used one is study guides including books that are detailed and suitable for building knowledge from the ground up and the other one is video tutorial and lectures that can somehow ease the pain of through study and are comparatively less boring for some candidates. learning from both videos and books needs lots of concentration and commitment. Along with going through all the defined course contents, participants should include practice exams. Statistics have shown that most students fail because of not having proper guidance. Certification-questions.com expert team recommends you to prepare some notes on these topics along with it don't forget to practice Adobe **ADO-E200 dumps** which has been written by our expert team, Both these will help you a lot to clear this exam with good marks.

How much ADO-E200 Exam Cost:

The price of ADO-E200 exam is 180 USD.

How to book the ADO-E200 Exam:

There are the following steps for registering the Adobe ADO-E200 exam:

- Step 1: Visit Adobe's credential management system logon page. The logon requires that you have an Adobe ID. If you do not have an Adobe ID, click the link for "Get an Adobe ID".
- Step 2: After logging on, if you have never taken an Adobe exam before, you will be instructed to create a Profile. After creating your profile, you will be instructed to update your opt-in Settings.
- Step 3: Once logged on, click "Home" then click "Schedule your exam at PSI."
- Step 4: You will be directed to a new page within CertMetrics where you will click "Click here to log in to PSI".
- Step 5: You will be taken to a page hosted by our exam delivery vendor, PSI, that displays the available exams.
- Step 6: Scroll through the list of available exams and press the "Schedule Exam" button for the exam you want to take.
- Step 7: Select a delivery mode for your exam by choosing either "Test Center" or "Remote Online Proctored Exam."
- Step 8: Select your exam language to see options for your exam.
- Step 9: Select an available date and start time.

- Step 10: Confirm schedule details to receive a booking confirmation.
- Step 11: Proceed to payment.
- Step 12: After payment is successful, you will receive an email confirmation with registration details and a receipt.

What is the duration, language, and format of ADO-E200 Exam:

- Format: Multiple choices, multiple answers
- Length of Examination: 120 minutes
- Number of Questions: 70
- Passing score: 67.14%
- language: English

Salary of Adobe Analytics Architect ADO-E200:

The Average Salary of an Adobe Analytics Architect-

- United State - 91292 USD
- India - 919078 INR
- Europe - 60198 EURO
- England - 62807 POUND

The benefit in Obtaining the ADO-E200 Exam Certification:

- AD0-E200 Certified candidates use to have a digital badge from Adobe and that badge they can place it on their resume.

- AD0-E200 certified candidates will be confident and stand different from others as their skills are more trained than non-certified professionals.

- Adobe **AD0-E200 Exam** will provide proven knowledge to use the tools to complete the task efficiently and effectively.

- AD0-E200 Certification provides practical experience to candidates from all the aspects to be a proficient worker in the company.

- AD0-E200 Certifications will provide opportunities to get a job easily as compared to non-certified individuals.

- AD0-E200 certified individuals get higher earning potential and increased promotion opportunities.

Difficulty in writing ADO-E200 Exam:

Many Candidates use to face many problems when they start preparing for the Adobe ADO-E200 exam. They use to have fear in their mind about the exam like how questions would be coming and how easily they can crack this exam. If a candidate wants to prepare for the Adobe ADO-E200 exam without facing any major challenge and willing to secure a good grade in the exam. Then they have to choose the best Adobe ADO-E200 dumps which should have real exam questions practice. Many websites are offering the latest Adobe ADO-E200 exam questions and answers but these questions are not verified by Adobe certified experts and that's why many participants use to get fail in their just first attempt. Certification-questions are the best platform that provides the candidate with the necessary Adobe ADO-E200 questions that will help him to pass the Adobe ADO-E200 exam. Adobe **ADO-E200 dumps** provided by the Certification-questions, Candidate will have every valuable material required to pass the Adobe ADO-E200 exam. We are providing the latest and actual questions and that is the reason why candidates need to use our provided study guide and exam dumps. We have the guarantee that you can expect almost the same questions in the exam which we are providing in our Adobe ADO-E200 dumps which would be helping you to clear Adobe ADO-E200 Exam in the first attempt itself.

For more info visit::

Adobe ADO-E200 Exam Guide
Adobe ADO-E200 Preparation Guide
Adobe Learn and Support

Sample Practice Test for AD0-E200

Question: 1 *One Answer Is Right*

An e-commerce website has recorded a revenue value of "100" with currency='EUR set on the page load. The admin of Report suite has set the base currency of INR. Which value will be shown in the reports?

Answers:

A) INR: 100 will be converted into EUR then shown in the reports.

B) EUR e100 will be converted into INR then shown in the reports

C) EUR e100 will be shown

D) INR:100 will be shown

Solution: C

Question: 2 *One Answer Is Right*

Which of the following cannot be used in a conversion funnel report?

Answers:

A) Conversion variables

B) Custom Events

C) Calculated Metric

D) Segments

Solution: A

Question: 3 *One Answer Is Right*

After implementing the percentPageViewed plugin which captures how much of the previous page was viewed you decided to group the values into more manageable business for cleaner reporting. After evaluating your option you decide to create a regulate expression that will automatically group values into 25% buckets. Based on this information, which regular expression can correctly classify data for the 51+ 75 bucket?

[5][1-9]+[6][0-9]+[7][0-5]

[51-59]+[60-69]+[70-75]

[51-59][[60-69]][70-75]

[5][1-9][[6][0-9][[7][0-5]

[5][1-9]&[6][0-9]&[7][0-5]

Answers:

A) Option A

B) Option B

C) Option C

D) Option D

Solution: A

Question: 4 *Multiple Answers Are Right*

In chronological order from 1 (top) to 4 (bottom), what are the steps you will perform to make positive changes to a new website? Select the correct answer for each step from the dropdown list.

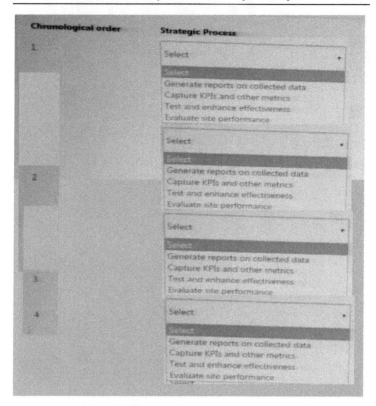

Answers:

A)

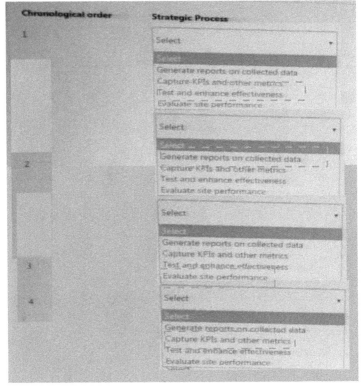

Solution: A

Explanation:

Explanation:

Question: 5 *One Answer Is Right*

While reviewing data for an existing site, you notice that all search terms captured in a custom variable during a visit credit toward success events. Which of the following topics could explain this behavior? Select two.

Answers:

A) Last Touch Allocation

B) Participation

C) Serialization

D) Linear Allocation

E) Merchandising

Solution: A, D

Question: 6 *One Answer Is Right*

You are being asked to help someone in your organization who is having difficulty accessing Adobe Launching using the Experience Cloud. Upon clicking the link to access Adobe Launch they are taken to the main Properties tab as expected. However they are not able to access the necessary web property that they need in order to complete the needed work. Which option below would be the best solution?

Answers:

A) Add the appropriate Adobe Launch Product Profile to the user account

B) An authorized client care user in your organization must contact Adobe to get the issue resolved

C) Create an Adobe Launch Administrator account for the user

D) Add the appropriate Adobe Launch Product assigned to their user account

Solution: D

Question: 7 *One Answer Is Right*

An analyst is copying settings from an report suite to a new report suite. Which settings are copied over to a new report suite? Select two.

Answers:

A) Evar

B) VISTA rules

C) Enable transaction ID Recording

D) Pathing on variable

E) Classification

Solution: B, E

Question: 8 *One Answer Is Right*

A marketing decides to track search refinement that were selected in the search results page. A list was implemented to capture filters in propel using as the delimiter. Suppose a visitor made 10 search result pate view consistent refinement of sort1/sort3'. What will be page view count recorded in the prop1 report?

Answers:

A) 10 page views to sort3

B) 10 page views to sort/sort2/sort3

C) 10 page views to sort1

D) 10 page views to sort2

E) 10 page views each to sirt1.sort2. sort3

Solution: B

Question: 9 *One Answer Is Right*

While reviewing analytics report for your website one day, you open up to the exit links report and notice it is empty. After manually testing some exist links on your site. You do not see the any image requests firing for external links either. What could be causing the missing images for external links? Select two.

Answers:

A) s.linkInternalFilters has not been property defined

B) s.trackExternalLinks is set to false

C) s.trackLink has not been properly defined

D) s.trackDownloadlinks is set to true

E) s.link tracVar is set to false

Solution: A, E

Question: 10 *One Answer Is Right*

Which settings determiners how long Adobe will a custom eVar value?

Answers:

A) Allocation

B) Classification

C) Merchandising

D) expiration

E) Correlation

Solution: C

Chapter 19: AD0-E201 - Adobe Analytics Developer

Exam Guide

How to Prepare For Adobe Analytics Developer ADO-E201 Exam

Preparation Guide for Adobe Analytics Developer ADO-E201 Exam

Introduction

Adobe certification is the swift, economical, and official way to demonstrate your skills using a range of industry-standard software packages, so it's probably a good investment. You don't need to have training in these tools before taking the exam, but it is extremely advisable to have proper experience of the required Adobe Tools. Adobe's certification exams are developed following industry-accepted standards to ensure validity and reliability. We work with industry experts to create our exams, which represent real-world requirements and objectives for the job roles we certify.

Adobe Analytics Developer ADO-E201 Exam:

Adobe Analytics Developer **ADO-E201 Exam** is related to Adobe Certified Expert Certification. This exam validates the Candidate knowledge and skills of adobe analytics development as per the business needs. It helps to demonstrate solution Design Reference to develop efficient code for data capture.

ADO-E201 Exam topics:

Candidates must know the exam topics before they start of preparation. Our Adobe **ADO-E201 dumps** will include the following topics:

1: Understanding Analytics in the Adobe Experience Cloud Eco-system 5%

- Demonstrate knowledge on utilizing the Adobe Experience Cloud ID
- Demonstrate understanding of triggers and activation
- Demonstrate understanding of common integrations across the Adobe Experience Cloud Eco-system

2: Analytics Strategy and Design based on a Solution Design Reference 10%

- Interpret a Solution Design Reference to develop efficient code for data capture
- Use the Tech Spec for populating data objects

3: Analytics implementation and configuration 30%

- Determine which environment to deploy Adobe Analytics Code
- Utilize the variables required for a base code Analytics configuration
- Apply the process to configure Adobe Analytics non-report suite settings
- Apply the process to configure Adobe Analytics report suite and variable settings
- Apply the process to utilize the data layer within a deployment

4: Tag Management Systems 25%

- Enumerate requirements to deploy Adobe Experience Platform Launch in Adobe Analytics
- Apply processes to configure website tagging with Adobe Launch
- Enumerate requirements to enable extensions and Adobe Launch

   5: Mobile Services 13%

- Apply processes to configure and execute a base predeployment code for mobile
- Demonstrate understanding of data dictionaries and configure processing rules for mobile

   6: Troubleshooting, testing and validating 17%

- Troubleshoot JavaScript errors
- Troubleshoot Adobe Analytics variables
- Recommend testing and debugging best practices for a given scenario
- Troubleshoot an Adobe Analytics server call across its lifecycle

   Certification Path:

There is no prerequisite for this Adobe ADO-E201 exam.

Who should take the ADO-E201 exam:

ADO-E201 exam focus more on Architects, Developers, Implementation Specialists/Engineers. Those candidates who are planning to have it should have a minimum level of knowledge or experience:

- Front end web development (e.g. JavaScript, CSS, HTML, jQuery Selector, basic RegEx)

- 1 - 2 years of experience deploying and configuring Adobe Analytics
- Ability to read and interpret a solution design document
- Familiarity with input/output methods (e.g. data feeds, data sources, APIs) and data extension methods
- overs classifications, mapping, etc.)
- At least 6 months of Analytics reporting experience
- Understanding of mobile SDKs, and corresponding features
- Experience with Q/A testing processes and tools
- Understands Adobe (Experience Platform) Launch and migrating DTM to Launch
- 6+ months experience deploying tag management systems including performance implications
- dobe DTM or Launch)
- Understanding of common data layer methodology
- Able to design a data model (data layer) and provide a cohesive approach

How to study the ADO-E201 Exam:

For Adobe Exam it is really important for any participant that they should have a proper study plan in place because simply reading without proper planning will not be adding any values. There are many sources are the there on the internet, mainly two types of resource can be used one is study guides including books that are detailed and suitable for building knowledge from the ground up and the other one is video tutorial and lectures that can somehow ease the pain of through study and are comparatively less boring for some candidates. learning from both videos and books needs lots of concentration and commitment. Along with going through all the defined course contents, participants should include practice exams. Statistics have shown that most students fail because of not having proper guidance. Certification-questions.com expert team recommends you to prepare some notes on these topics along with it don't forget to practice Adobe **ADO-E201 dumps** which has been written by our expert team, Both these will help you a lot to clear this exam with good marks.

How much ADO-E201 Exam Cost:

The price of ADO-E201 exam is 180 USD.

How to book the ADO-E201 Exam:

There are the following steps for registering the Adobe ADO-E201 exam:

- Step 1: Visit Adobe's credential management system logon page. The logon requires that you have an Adobe ID. If you do not have an Adobe ID, click the link for "Get an Adobe ID".

- Step 2: After logging on, if you have never taken an Adobe exam before, you will be instructed to create a Profile. After creating your profile, you will be instructed to update your opt-in Settings.

- Step 3: Once logged on, click "Home" then click "Schedule your exam at PSI."

- Step 4: You will be directed to a new page within CertMetrics where you will click "Click here to log in to PSI".

- Step 5: You will be taken to a page hosted by our exam delivery vendor, PSI, that displays the available exams.

- Step 6: Scroll through the list of available exams and press the "Schedule Exam" button for the exam you want to take.

- Step 7: Select a delivery mode for your exam by choosing either "Test Center" or "Remote Online Proctored Exam."

- Step 8: Select your exam language to see options for your exam.
- Step 9: Select an available date and start time.
- Step 10: Confirm schedule details to receive a booking confirmation.
- Step 11: Proceed to payment.
- Step 12: After payment is successful, you will receive an email confirmation with registration details and a receipt.

What is the duration, language, and format of ADO-E201 Exam:

- Format: Multiple choices, multiple answers
- Length of Examination: 120 minutes
- Number of Questions: 60
- Passing score: 68.33%
- language: English

Salary of Adobe Analytics Developer ADO-E201:

The Average Salary of an Adobe Analytics Developer-

- United State - 90291 USD
- India - 149078 INR
- Europe - 65198 EURO
- England - 72807 POUND

The benefit in Obtaining the ADO-E201 Exam Certification:

- AD0-E201 Certified candidates use to have a digital badge from Adobe and that digital badge can attract more recruiters when they will place it on their CV.

- AD0-E201 certified candidates will be confident and stand different from others as they having more skills in terms of delivery of the project.

- Adobe **AD0-E201 Exam** will provide proven knowledge to use the tools to complete the task efficiently and effectively.

- AD0-E201 Certification provides practical experience to candidates from all the aspects to be a proficient worker in the company.

- AD0-E201 Certifications will provide opportunities to get a job easily as compared to non-certified individuals.

- AD0-E201 certified individuals get higher earning potential and increased promotion opportunities.

- AD0-E201 credential delivers higher earning potential and increased promotion opportunities because it shows a good understanding of manage topologies cadences.

Difficulty in writing ADO-E201 Exam:

Many Candidates use to face many problems when they start preparing for the Adobe ADO-E201 exam. They use to have fear in their mind about the exam like how questions would be coming and how easily they can crack this exam. If a candidate wants to prepare for the Adobe ADO-E201 exam without facing any major challenge and willing to secure a good grade in the exam. Then they have to choose the best Adobe ADO-E201 dumps which should have real exam questions practice. Many websites are offering the latest Adobe ADO-E201 exam questions and answers but these questions are not verified by Adobe certified experts and that's why many participants use to get fail in their just first attempt. Certification-questions is the best platform which provides the candidate with the necessary

Adobe ADO-E201 questions that will help him to pass the Adobe ADO-E201 exam. **Adobe ADO-E201 dumps** provided by the Certification-questions, Candidate will have every valuable material required to pass the Adobe ADO-E201 exam. We are providing the latest and actual questions and that is the reason why candidates need to use our provided study guide and exam dumps. We have the guarantee that you can expect almost the same questions in the exam which we are providing in our Adobe ADO-E201 dumps which would be helping you to clear Adobe ADO-E201 Exam in the first attempt itself.

For more info visit::

Adobe ADO-E201 Exam Guide
Adobe ADO-E201 Preparation Guide
Adobe ADO-E201 Sample Test
Adobe Learn and Support

Sample Practice Test for AD0-E201

Question: 1 *One Answer Is Right*

A developer needs to have a rule fire as late as possible upon the initial view of the page. Which event type is the latest to load?

Answers:

A) DOM Ready

B) Window Loaded

C) Library Loaded

D) Page Bottom

Solution: B

Question: 2 *One Answer Is Right*

A client has a hybrid native app that opens mobile web content the app. Which method should be used to ensure that visitors are not identified separateltly as they move between the native and mobile web pages?

Answers:

A) VisitorMarkitingCloudID()

B) VisitorSyncIdentifiers()

C) visitorAppendTourL()

D) visitorgetIDs

Solution: C

Question: 3 *One Answer Is Right*

When adding a new app in the Adobe Mobile Services interface, which three settings are required to save and create the app? (Choose three.)

Answers:

A) Name

B) Use HTTPS

C) App Store

D) Type

E) Report Suite

F) Organization

Solution: A, D, E

Question: 4 *One Answer Is Right*

What is the first step needed in order to configure a mobile application that will implement tracking using Adobe Experience Platform Launch?

Answers:

A) Check the "enable mobile tracking" option in the property configuration in Launch.

B) Add the Mobile Application extension to the property in Launch.

C) Configure the SDK with Environment ID from Launch.

D) Add the Launch embed code into the application.

Solution: C

Question: 5 *One Answer Is Right*

Some variables are being populated correctly in you implementation and others are not Which of the displayed variables uses the correct case?

Answers:

A) s.channel-Help

B) s.PageType-Cprt'

C) s.Pagename-'Homepage

D) s.purchased-1234567899'

Solution: A

Question: 6 *One Answer Is Right*

Your is the contains many links to external sites. You do not want to track all exist liks, but only report on a specific subnet of them. Adobe Analytics has the following configuration:

```
s.trackExternalLinks=true;
s.linkInternalFilters="javascript:,mysite.com";
s.linkExternalFilters="site1.com,site2.com,site3.com/partners";
s.linkLeaveQueryString=false;
```

Which two links are considered external links? (Choose two.) A)

```
&lt;a href="https://www.mysite.com"&gt;Home&lt;/a&gt;
```

B)

```
&lt;a href="https://www.site3.com"&gt;Partner Portal&lt;/a&gt;
```

C)

```
&lt;a href="https://www.site1.com"&gt;Follow us&lt;/a&gt;
```

D)

```
&lt;a href="https://www.site3.com/partners/offer.html"&gt;Offers&lt;/a&gt;
```

E)

```
&lt;a href="/careers/job_list.html"&gt;Careers&lt;/a&gt;
```

Answers:

A) Option A

B) Option B

C) Option C

D) Option D

Solution: C, D

Question: 7 *One Answer Is Right*

Ah Adobe Analytics report shows a big percentage of breakdown values as "Other". What could be the reason for this problem?

Answers:

A) Conversion event fires without a conversion variable

B) Using segments in which in the rule is disabled

C) Pages fire outside URL filters

D) Unclassified data in classification reports

Solution: C

Question: 8 *One Answer Is Right*

An adobe Analytics developer is analyzing the reasons behind a data loss. The application has to support offline tracking, so trackoffline is set to true. What is the reason behind the data loss?

Answers:

A) Report is not timestamp enabled.

B) Default value for property offliceLimit is not increased.

C) Forceoffline property is enabled.

D) Additional property offlinetrackingSuite is empty.

Solution: A

Question: 9 *One Answer Is Right*

While navigating from one page to another, two server calls are sent with each click. The one has a link type pf Ink_e"\ which is used only for Exist Link Which variable should be modified to resolved the issue with the extra server call?

Answers:

A) s, linkInternalFilters

B) s,trackExiyLinks

C) s.linkExternalFilters

D) s.trackinternall;inks

Solution: A

Question: 10 *One Answer Is Right*

An Analytics developer placed the doPlugins() function in the code editor of the Adobe Analytics extension to populate the s.campaign variable. Then, the developer created a rule to populated the s,compaign variable on a specific landing page. When validating the value of the s,compaign variable on the specific landing page, it shown the doPlugins() value for the campaign variable. What would cause this issue?

Answers:

A) The doPlugin() function value will take precedence over the s,compaign variable

B) The doPlugin() function needs to be excluded through an exception in the rule.

C) The doPlugin() function runs last and overwrites any values for the same variables.

D) The rules condition was not met. so the doPlugins() function persisted.

Solution: C

Chapter 20: AD0-E202 - Adobe Analytics Business Practitioner

Exam Guide

How to Prepare For Adobe Analytics Business Practitioner AD0-E202 Exam

Preparation Guide for Adobe Analytics Business Practitioner AD0-E202 Exam

Introduction

Adobe certification is the swift, economical, and official way to demonstrate your skills using a range of industry-standard software packages, so it's probably a good investment. You don't need to have training in these tools before taking the exam, but it is extremely advisable to have proper experience of the required Adobe Tools. Adobe's certification exams are developed following industry-accepted standards to ensure validity and reliability. We work with industry experts to create our exams, which represent real-world requirements and objectives for the job roles we certify.

Adobe Analytics Business Practitioner AD0-E202 Exam:

Adobe Analytics Business Practitioner **AD0-E202 Exam** is related to Adobe Certified Expert Certification. This exam validates the Candidate knowledge and skills of digital marketing, digital analytics, understands the business value of web technologies.

AD0-E202 Exam topics:

Candidates must know the exam topics before they start of preparation. Because it will help them in hitting the core. Our Adobe **AD0-E202 dumps** will include the following topics:

1: Business Analysis (25%)

- Given a business need/question, identify the most appropriate reporting strategy to perform an analysis
- Analyze data to answer business questions and recommend new optimization hypotheses
- Identify conversion funnels
- Interpret Solution Design Reference (SDR) to determine what data is available in reports
- Analyze report data to summarize and draw conclusions
- Understand outliers and anomalies in reports

   2: Reporting and Dashboarding for Projects (25%)

- Apply the process to configure projects using the most appropriate tool(s)
- Compare fallout and flow visualization and appropriate variable types for reporting
- Apply the process to schedule Projects, Report Builder and Data Warehouse
- Apply the process to share Projects and Reporting and Analytics dashboards for different users and/or groups
- Apply the process to set Alerts
- Apply the process to lookup the dimensions/components
- Apply the process to create a visualization
- Given a scenario, determine the appropriate item to use

   3: Segmentation and Calculated Metrics (25%)

- Determine how to develop and configure segments

- Apply the process to share segments with others in the organization

- Compare segments

- Apply segments to Projects and Components

- Apply the process to generate calculated and/or segmented metrics

- Determine Metric Types
  4: General Tool Knowledge and Troubleshooting (15%)

- Analyze reports and identify data quality issues

- Define different types of dimensions and parameters existing in Adobe Analytics

- Determine how to bring data in and out of Adobe Analytics

- Apply the process to configure Report Builder

  5: Administration (10%)

- Apply the process to configure the Marketing Channel reports with Marketing Channel processing rules
- Apply the process to configure Classification Importer and Rule Builder
- Identify information from marketing URLs
- Apply the process to configure a virtual report suite based upon an existing segment
- Understand Report Suite Admin Console Settings

  Certification Path:

There is no prerequisite for this Adobe AD0-E202 exam.

Who should take the AD0-E202 exam:

AD0-E202 exam has been designed for the following Target Audience:-

- Digital marketers
- Business analysts
- Business consultants
- Data analysts
- Web analysts
- Digital analysts
- Media/marketing analysts
- Product owners and managers
- UI analysts
- Conversion/optimization specialists

There is a minimum skill set required before appearing for **AD0-E202 exam**:

- Has 2 years' experience in digital analytics, typically has a marketing background or comes from a marketing role but can come from a variety of job roles, possesses an in-depth understanding of digital analytics metrics and dimensions, understands the business value of web technologies, can translate business requirements into metrics or KPIs, build and interpret reports, communicate results and can propose a course of action based on analysis of reports
- Helps clients understand how to extract/pull the information they want, and which are contextually relevant (e.g. help them define their business questions)
- Understands how the technology works and understands a client's implementation (e.g. where the data sits, how it can be collected, how it is tracked) from a functional standpoint
- Has at least one year of Adobe Analytics hands-on experience

- Has a basic understanding of how digital analytics is filtering/structuring data
- Has an awareness of the integration of Adobe Analytics with other Adobe solutions

How to study the AD0-E202 Exam:

For Adobe Exam it is really important for any participant that they should have a proper study plan in place because simply reading without proper planning will not be adding any values. There are many sources are the there on the internet, mainly two types of resource can be used one is study guides including books that are detailed and suitable for building knowledge from the ground up and the other one is video tutorial and lectures that can somehow ease the pain of through study and are comparatively less boring for some candidates. learning from both videos and books needs lots of concentration and commitment. Along with going through all the defined course contents, participants should include practice exams. Statistics have shown that most students fail because of not having proper guidance. Certification-questions.com expert team recommends you to prepare some notes on these topics along with it don't forget to practice Adobe **AD0-E202 dumps** which has been written by our expert team, Both these will help you a lot to clear this exam with good marks.

How much AD0-E202 Exam Cost:

The price of AD0-E202 exam is 180 USD.

How to book the AD0-E202 Exam:

There are the following steps for registering the Adobe AD0-E202 exam:

- Step 1: Visit Adobe's credential management system logon page. The logon requires that you have an Adobe ID. If you do not have an Adobe ID, click the link for "Get an Adobe ID".

- Step 2: After logging on, if you have never taken an Adobe exam before, you will be instructed to create a Profile. After creating your profile, you will be instructed to update your opt-in Settings.
- Step 3: Once logged on, click "Home" then click "Schedule your exam at PSI."
- Step 4: You will be directed to a new page within CertMetrics where you will click "Click here to log in to PSI".
- Step 5: You will be taken to a page hosted by our exam delivery vendor, PSI, that displays the available exams.
- Step 6: Scroll through the list of available exams and press the "Schedule Exam" button for the exam you want to take.
- Step 7: Select a delivery mode for your exam by choosing either "Test Center" or "Remote Online Proctored Exam."
- Step 8: Select your exam language to see options for your exam.
- Step 9: Select an available date and start time.
- Step 10: Confirm schedule details to receive a booking confirmation.
- Step 11: Proceed to payment.
- Step 12: After payment is successful, you will receive an email confirmation with registration details and a receipt.

What is the duration, language, and format of the AD0-E202 Exam:

- Format: Multiple choices, multiple answers
- Length of Examination: 120 minutes
- Number of Questions: 69
- Passing score: 69%

- language: English and Japanese

    Salary of Adobe Analytics Business Practitioner:

The Average Salary of an Adobe Analytics Business Practitioner in

- United State - 71292 USD
- India - 819078 INR
- Europe - 50198 EURO
- England - 52807 POUND

    The benefit in Obtaining the AD0-E202 Exam Certification:

- Resumes with Adobe Certified Expert certifications get noticed and fast-tracked by hiring managers.

- AD0-E202 Certified candidates use to have a digital badge from Adobe and that digital badge can attract more recruiters when they will place it on their CV.

- AD0-E202 certified candidates will be confident and stand different from others as they having more skills in terms of delivery of the project.

- Adobe **AD0-E202 Exam** will provide proven knowledge to use the tools to complete the task efficiently and effectively.

- AD0-E202 Certification provides practical experience to candidates from all the aspects to be a proficient worker in the company.

- AD0-E202 Certifications will provide opportunities to get a job easily as compare to non certified individuals.

- AD0-E202 certified individuals get higher earning potential and increased promotion opportunities.

- AD0-E202 credential delivers higher earning potential and increased promotion opportunities because it shows a good understanding of manage topologies cadences.

Difficulty in writing AD0-E202 Exam:

Adobe AD0-E202 is one of the advanced level certifications from Adobe, Many Candidates use to face many problems when they start preparing for the Adobe AD0-E202 exam. They use to have fear in their mind about the exam like how questions would be coming and how easily they can crack this exam. If a candidate wants to prepare for the Adobe AD0-E202 exam without facing any major challenge and willing to secure a good grade in the exam. Then they have to choose the best Adobe AD0-E202 dumps which should have real exam questions practice. Many websites are offering the latest Adobe AD0-E202 exam questions and answers but these questions are not verified by Adobe certified experts and that's why many participants use to get fail in their just first attempt. Certification-questions is the best platform which provides the candidate with the necessary Adobe AD0-E202 questions that will help him to pass the Adobe AD0-E202 exam. **Adobe AD0-E202 dumps** provided by the Certification-questions, Candidate will have every valuable material required to pass the Adobe AD0-E202 exam. We are providing the latest and actual questions and that is the reason why candidates need to use our provided study guide and exam dumps. We have the guarantee that you can expect almost the same questions in the exam which we are providing in our Adobe AD0-E202 dumps which would be helping you to clear Adobe AD0-E202 Exam in the first attempt itself.

For more info visit::

Adobe AD0-E202 Exam Reference

Adobe Learn and Support

Adobe ADO-E202 Sample Test Paper

Sample Practice Test for AD0-E202

Question: 1 *One Answer Is Right*

The product team wants to upload product pricing data from an offline database to the transaction ID in Adobe Analytics. Which data import tool should be used for this task?

Answers:

A) Data sources

B) Data Connecter

C) Classification Importer

D) Adobe fastETL

Solution: A

Explanation:

Explanation: When there is offline data you want permanently written into Adobe Analytics Options: Summary: simple data uploads, by day or limited dimensions Transaction ID: data uploads that connect an online endpoint to offline data, and fully associate imported data to a visitor snapshot captured online (e.g. orders complete online, and get returned offline)

Transaction ID data sources allow you to not only view online and offline data side-by-side, but tie the data together. It requires the use of the transactionID variable in your Analytics implementation. When you send an online hit that contains a transactionID value, Adobe takes a "snapshot" of all variables set or persisted at that time. If a matching transaction ID uploaded through Data Sources is found, the offline and online data is tied together. It does not matter which source of data is seen first.

Question: 2 *One Answer Is Right*

Various reports show None Unspecified other, or unknow depending on the specific report viewed, generally, this breakdown means that the variable was not defined or otherwise unavailable. Which statements explains the possible behavior of the data?

Answers:

A) Similar to events firing eVars, It is possible to see "OTHER" in a merchandising eVar report when that variable is not defined before a success event.

B) Similarly to non-mobile hits in mobile reports, mobile hits in all Visitor profile| technology report when that variable is not defined before a success event.

C) When viewing classification data any value that does not have data associated with that particular returns "NONE". To resolve this issue create a classification export file and classify the appropriate columns.

D) This happens when a user comes to your site for the first and males a purchase without firing eVar1. If you view order in the eVar1 report, there is no value to attribute this order to, so it will appear as NONE.

Solution: B

Question: 3 *One Answer Is Right*

An analyst has to alter a request in report Builder in order to illustrate the number of order by month and by purchase country. Currently it displays only the total values by month. What are two ways the analyst can address this change? (Choose two.)

Answers:

A) Edit the existing request and select the Dimension "Country" in the first step of the request wizard and "Month" in the second step of the request wizard.

B) Right-click on the request cell, and added depended Request > breakdown.

C) Right-click on the request and add Matching request >Breakdown.

D) Edit the existing request and select the dimension "Month" in the first step of the request wizard and Country" in the second step of the request wizard.

E) Edit the existing request, and select the Dimension "Month' and" Country" in the first step of the request wizard

Solution: A, D

Question: 4 *One Answer Is Right*

An Analyst need to create a report using a mix of pages viewed by visitors and customer events to understand where visitors abandon the process. Which type of report should the analyst create?

Answers:

A) A patting report within Report

B) A Pathfinder report within workspace

C) A fallout analyst within Workspace

D) A custom event funnel within Reports

Solution: B

Question: 5 *One Answer Is Right*

Analyzing the data in the image below, a data analyst verifies that the report of "entries" per page, shows that the HomePage A" page is responsible for only 3.4% of the total entries to the site:

Analyzing the data in the image below. A data analyst verifies that the report of "entries" per page, shows that the HomePage A" page responsible for only 4.4% of the total entries to the site:

Based on both images and knowledge on the standard metrics of Adobe Analytics, which statement is correct?

Answers:

A) The page Dimension should be used exclusively with the metric of page Views avoiding false analysis.

B) Occurrences refers to all hits associated with a particular entry page, also counting events Triggered throughout the session.

C) The entry page Dimension refers exclusively to the first page that a visitor lands on the site throughout its history, so there are distortions regarding the period of analysis.

D) The entries metric refers to entries on a given page, and is counted every time the page loads on a visit.

Solution: C

Question: 6 *One Answer Is Right*

Which option should a Shared User select In the Dashboard Manager to see changes/Updates made by the Dashboard Owner?

Answers:

A) On Menu

B) Copy Me

C) Duplicate Me

D) Dashboard Player

Solution: A

Explanation:

Explanation:
https://docs.adobe.com/content/help/en/analytics/analyze/re
ports-analytics/dashboard- manage.html

Question: 7 *One Answer Is Right*

A client requests a donut chart of the top 10 cities. The analyst
creates the following Freeform table.

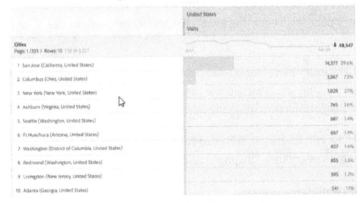

The following donut chart was already created, but only shows
five cities and an "Other" bucket.

What is the correct way to show all 10 cities in the same donut
visualization?

Answers:

A) Open the visualization settings in the top right visualization
menu, and set the item limit to 10.

B) Right-click the visualization to set the item limit to 10.

C) Change the source freeform table to limit the amount of records.

D) Click on the "Other" bucket label to unhide the list of elements in this bucket.

Solution: A

Question: 8 *One Answer Is Right*

A web analyst is viewing a report in an Analytics' Workspace freeform table and sees a dark grey triangle.

What is the specifying in the report?

Answers:

A) Incomplete data

B) End of data set

C) Data anomaly detected

D) Data includes outside data source

Solution: C

Question: 9 *One Answer Is Right*

The investment bank of a co-operative users Adobe analytics to monitor the use of its online banking. They use it to measure Appointment scheduled with account managers, who will then call clients al specified times. The IT team decides to integrate the appointment data captured in Adobe Analytics with the manager's calendar, as close to real-time as possible. How can this need to meet?

Answers:

A) Create a project in Analytics Workspace with broken data per hour and a line graph per minute. So you can check the number of schedules.

B) Use the Adobe Analytics API to have the Hit data in the shortest possible time, and thus cross the agenda of the bank account managers.

C) Use an hourly time export from Adobe Analytics Data Warehouse to integrate data with the account manager's schedule data.

D) Perform the extraction hourly via Data Feed, and cross them through the Data Stamp with the data of the agents of the account manager backing an hour.

Solution: C

Question: 10 *One Answer Is Right*

The e-commerce team of a home furniture store notices a decrease in numbers of orders over the last two months for the bedroom category. Which three metrics are relevant to include in the report to investigate the decrease? Fallout rate for each step in the purchase funnel for the bedroom" category

Answers:

A) Cart additional for bedroom" category

B) Revenue for the bathroom" category

C) Page views for bathroom" category

D) Average Page views per category

E) Orders per product in the bedroom category

Solution: A, D, E

Chapter 21: AD0-E301 - Campaign Standard Developer

Exam Guide

How to Prepare For Adobe Campaign Standard Developer ACE AD0-E301 Exam

Preparation Guide for Adobe Campaign Standard Developer ACE AD0-E301 Exam

Introduction

Adobe certification are swift, economical, and official way to demonstrate your skills using a range of industry-standard software packages, so it's probably a good investment. You don't need to have training in these tools before taking the exam, but it is extremely advisable.

Adobe Campaign Standard Developer ACE AD0-E301 Exam:

Adobe Campaign Standard Developer ACE **Adobe AD0-E301 Exam** is related to Adobe Campaign Standard Developer ACE and Credits toward Adobe Campaign Standard Developer Certification. This exam validates the ability to understand Adobe experience cloud web environment, ACS implementation, topology rules creation, transactional messaging, customizing workflows, minimal notions of RDBMS concepts, minimal notions of javascript, WYSWIG editors. Browser functionality, email processes and query tool knowledge is verified through this accreditation . Experience in cloud admin console / adobe admin console will be helpful in performing Adobe Campaign Standard Developer **Adobe AD0-E301 Exam**

Topics of Adobe Campaign Standard Developer ACE AD0-E301 Exam:

Candidates must know the exam topics before they start of preparation. Because it will really help them in hitting the core. Our **Adobe AD0-E301 dumps** will include the following topics:

1. Adobe Campaign Standard Solution 13%

- Enumerate ACS Concepts and Capabilities

    2. Data Modeling and Data Management 28%

- Extend the data resource
- Configure settings for data viewing and filtering
- Manage data in ACS

    3. Campaign Setup 28%

- Use templates
- Configure workflows
- Configure landing pages
- Configure profiles and audiences
- Configure personalization
- Configure typology

    4. Administration 10%

- Configure organizational units
- Configure security
- Apply processes on packages
- Apply processes to execute a workflow

    5. System Configuration 10%

- Configure external accounts
- Describe resources for GDPR compliance

6. Monitoring 8%

- Enumerate solution monitoring options
- Interpret log files
- Explain the double opt in set up
- Configure notification settings

7. Reporting 3%

- Configure custom reports

Certification Path:

The Adobe Campaign Standard Developer ACE certification path includes only one AD0-E301 certification exam.

Who should take the AD0-E301 exam:

The Adobe Campaign Standard Developer ACE **Adobe AD0-E301 Exam** certification is an internationally-recognized validation that identifies persons who earn it as possessing skilled in Adobe Campaign Standard Developer . If a candidate wants significant improvement in career growth needs enhanced knowledge, skills, and talents. The Adobe Campaign Standard Developer ACE AD0-E301 Exam certification provides proof of this advanced knowledge and skill. If a candidate has knowledge of associated technologies and skills that are required to pass Adobe Campaign Standard Developer ACE AD0-E301 Exam then he should take this exam.

- Marketing Automation Administrator
- Adobe Experience cloud developer
- Email Marketing Analyst
- Digital Marketer

How to study the AD0-E301 Exam:

There are two main types of resources for preparation of certification exams first there are the study guides and the books that are detailed and suitable for building knowledge from ground up then there are video tutorial and lectures that can somehow ease the pain of through study and are comparatively less boring for some candidates yet these demand time and concentration from the learner. Smart Candidates who want to build a solid foundation in all exam topics and related technologies usually combine video lectures with study guides to reap the benefits of both but there is one crucial preparation tool as often overlooked by most candidates the practice exams. Practice exams are built to make students comfortable with the real exam environment. Statistics have shown that most students fail not due to that preparation but due to exam anxiety the fear of the unknown. Certification-questions.com expert team recommends you to prepare some notes on these topics along with it don't forget to practice AD0-E301 Exam dumps which been written by our expert team, Both these will help you a lot to clear this exam with good marks.

How much AD0-E301 Exam Cost:

The price of AD0-E301 exam is $180 USD.

How to book the AD0-E301 Exam:

Exams are delivered worldwide by PSI.

These are following steps for registering the AD0-E301 exam.

- Step 1: Visit Adobe's credential management system logon page. The logon requires that you have an Adobe ID. If you do not have an Adobe ID, click the link for "Get an Adobe ID"

- Step 2: After logging on, if you have never taken an Adobe exam before, you will be instructed to create a Profile.After creating your Profile, you will be instructed to update your opt-in Settings.

- Step 3: Once logged on, click "Home" then click "Schedule your exam at PSI."

- Step 4: You will be directed to a new page within CertMetrics where you will click "Click here to log in to PSI".

- Step 5: You will be taken to a page hosted by our exam delivery vendor, PSI, that displays the available exams

- Step 6: Scroll through the list of available exams and press the "Schedule Exam" button for the exam you want to take.

- Step 7: Select a delivery mode for your exam by choosing either "Test Center" or "Remote Online Proctored Exam."

- Step 8: Select your exam language to see options for your exam.

- Step 9: Select an available date and start time.

- Step 10: Confirm schedule details to receive a booking confirmation.

- Step 11: Proceed to payment.

- Step 12: After payment is successful, you will receive an email confirmation your registration details and a receipt.

What is the duration, language, and format of the AD0-E301 Exam:

- Format: Multiple choices, multiple select
- Length of Examination: 90 minutes
- Number of Questions: 50
- Passing Score: All adobe exams are reported on a scale of 300 to 700. The passing score is 550 . More details on scaled scoring

- Language: English
- Delivery: Online Proctored (required camera access) or test center proctored

The benefit in Obtaining the AD0-E301 Exam Certification:

- Resumes with Adobe Certified Expert certifications get noticed and fast-tracked by hiring managers.
- AD0-E301 Certification is distinguished among competitors. AD0-E301 certification can give them an edge at that time easily when candidates appear for employment interview, employers are very fascinated to note one thing that differentiates the individual from all other candidates.
- AD0-E301 certification will be more useful and relevant networks that help them in setting career goals for themselves. AD0-E301 networks provide them with the correct career guidance than non certified generally are unable to get.
- AD0-E301 certified candidates will be confident and stand different from others as their skills are more trained than non-certified professionals.
- **Adobe AD0-E301 Exam** will provide proven knowledge to use the tools to complete the task efficiently and cost effectively than the other non-certified professionals lack in doing so.
- AD0-E301 Certification provides practical experience to candidates from all the aspects to be a proficient worker in the organization.
- AD0-E301 Certifications will provide opportunities to get a job easily in which they are interested in instead of wasting years and ending without getting any experience.

- AD0-E301 credential delivers higher earning potential and increased promotion opportunities because it shows a good understanding of manage topologies cadences.

  Difficulty in writing AD0-E301 Exam:

you may encounter many problem when you start preparing for the AD0-E301 exam. These certification are usually tough and tricky. Most of the time it's been observed that exam dive into minutia, which are usually difficult for professionals even with years of experience. If a candidate wants to prepare for the AD0-E301 exam without any problem and get good grades in the exam. They should follow discipline such as organizing study places, taking proper breaks such as for every 1 hour, you study, take a short 10 minute break to recharge. Make studying less overwhelming by condensing notes from course.

Then they have to choose the best AD0-E301 dumps for real exam questions practice. There are many websites that are offering the latest AD0-E301 exam questions and answers but these questions are not verified by Adobe certified experts and that's why many are failed in their just first attempt. Certification-questions is the best platform which provides the candidate with the necessary AD0-E301 questions that will help him to pass the AD0-E301 exam on the first time. Candidate will not have to take the AD0-E301 exam twice because with the help of **Adobe AD0-E301 dumps** Candidate will have every valuable material required to pass the Adobe AD0-E301 exam. We are providing the latest and actual questions and that is the reason why this is the one that he needs to use and there are no chances to fail when a candidate will have valid dumps from Certification-questions. We have the guarantee that the questions that we have will be the ones that will pass candidate in the AD0-E301 exam in the very first attempt.

For more info visit::

Manage workflow in ACS Conversational marketing technology

Sample Practice Test for AD0-E301

Question: 1 *One Answer Is Right*

What is needed to verify that a landing page which is blacklisting an email channel for a profile, is working?

Answers:

A) A test profile subscribed to a service

B) A test profile

C) A profile

D) A blacklisted profile

Solution: C

Explanation:

Explanation: Reference:
https://docs.adobe.com/content/help/en/campaign-standard/using/profiles-and-audiences/ understanding-opt-in-and-opt-out-processes/about-opt-in-and-opt-out-in-campaign.html

Question: 2 *One Answer Is Right*

What are two steps a developer must do to set up a key-based authentication in order to transfer files to the Adobe-hosted SFTP? (Choose two.)

Answers:

A) Request Adobe support to whitelist IP addresses

B) Upload a private key in the external account configuration.

C) Upload a public key in the external account configuration.

D) Provide the public key to Adobe support to have it uploaded to the SFTP server.

E) Provide the private key to Adobe support to have it uploaded to the SFTP server.

Solution: A, D

Explanation:

Explanation: Reference:
https://docs.adobe.com/content/help/en/campaign-classic/using/getting-started/importing-and- exporting-data/sftp-server-usage.html

Question: 3 *One Answer Is Right*

Every week, new data files are uploaded to the Adobe-hosted SFTP server. What is the maximum time limit that files remain on the SFTP server?

Answers:

A) 30 days

B) 15 days

C) 20 days

D) 25 days

Solution: B

Explanation:

Explanation: Reference:
https://docs.adobe.com/content/help/en/campaign-classic/using/getting-started/importing-and- exporting-data/sftp-server-usage.html

Question: 4 *One Answer Is Right*

Which two component categories are used to filter data in a report? (Choose two.)

Answers:

A) Population

B) Metrics

C) Period

D) Segments

E) Targets

Solution: D, E

Question: 5 *One Answer Is Right*

For which reason would you configure a test profile for usage as a trap?

Answers:

A) to check the message before sending the finalized delivery

B) to preview a message to test the personalization elements

C) to check the way a message is displayed on a variety of email clients

D) to identify whether your client file is being used fraudulently

Solution: D

Explanation:

Explanation: Reference:
https://docs.adobe.com/content/help/en/campaign-standard/using/testing-and-sending/ preparing-and-testing-messages/managing-test-profiles-and-sending-proofs.html

Question: 6 *One Answer Is Right*

For what two types of data is it the best practice to export using the package export process? (Choose two.)

Answers:

A) Profiles

B) Content Templates

C) Logs

D) Business Data

E) Workflows

Solution: A, E

Explanation:

Explanation: Reference:
https://docs.adobe.com/content/help/en/campaign-classic/using/getting-started/importing-and- exporting-data/exporting-data.html

Question: 7 *One Answer Is Right*

Which configuration would allow a client to create a user that can access their French (FR) business unit but not allow them access to German (DE) campaigns?

Answers:

A) 1. Create new Security Groups named "FR Operators" and "DE Operators" 2. Map all FR marketing activities to "FR Operators" Security Group 3. Map the user to "FR Operators" group

B) 1. Create new Org Units named "FR" and "DE" 2. Create new Security Groups named "FR Operators" and "DE Operators" 3. Map Security groups to "FR" and "DE" Org Units, respectively 4. Map all FR marketing activities to "FR" Org Unit 5. Map the user to "FR Operators" group

C) 1. Create new Org Unit named "EU" 2. Create new Security Groups named "FR Operators" and "DE Operators" 3. Map Security groups to "EU" Org Unit 4. Map all marketing activities to "EU" Org Unit 5. Map the user to "FR Operators" group

D) 1. Create new Org Units named "FR" and "DE" 2. Map all FR marketing activities to "FR" Org Unit 3. Map the user to "FR" Org Unit

Solution: D

Question: 8 *One Answer Is Right*

Review the following error message: The schema for profiles specified in the transition ('head:cusEmployee') is not compatible with the schema defined in the delivery template ('nms:recipient'). They should be identical. What is causing this error?

Answers:

A) There is no reconciliation with profiles.

B) There is no audience set in the delivery.

C) The schema in the delivery template is incorrect.

D) The transition is not linked to the delivery.

Solution: C

Explanation:

Explanation: Reference:
https://forums.adobe.com/thread/2611457

Question: 9 *One Answer Is Right*

What does the package functionality allow you to do?

Answers:

A) It allows you to export only business data from one instance to another.

B) It allows you to import and export business data and configuration data from one instance to another.

C) It allows you to export only configuration data from one instance to another.

D) It allows you to create a full back-up of the database without database access.

Solution: B

Explanation:

Explanation: Reference:
https://helpx.adobe.com/campaign/kb/data-package-best-practices.html

Question: 10 *One Answer Is Right*

You want to export a file containing the Label value for all created emails. Which out-of-the-box resource do you need to query in the Export Activity?

Answers:

A) Campaign

B) Messages

C) Delivery

D) Logs

Solution: A

Explanation:

Explanation: Reference:
https://docs.adobe.com/content/help/en/campaign-classic/using/reporting/reporting-in-adobe- campaign/about-adobe-campaign-reporting-tools.html

SUMMARY

To recap, main stages of certification exam study guide are Introduction to Adobe Exam , Adobe Exam topics in which Candidates must know the exam topics before they start of preparation, Adobe Exam Requirements, Cost of Adobe Exam, registration procedure of the Adobe Exam, Adobe Exam formate, Adobe Exam Certified salary, AdobeExam advantages.

If you are aspirant to pass the cerification exam, start exam preparation with study material provided by Certification-questions.com

About The Author

David Mayer

Co-Founder of Certification-Questions.com

David is the Co-founder of Certification-Questions.com, one of the largest Certification practice tests and PDF exams websites on the Internet. They are providing dumps an innovative way by providing Online Web Simulator and Mobile App. He likes to share his knowledge and is active in the Adobe community.

He has written several books, blogs, and is active in the Adobe community.

APPENDIX

Certification

The action or process of providing someone or something with an official document that accredits a state or level of results.

Practice test

The practical exam is an alternative, non-scoring version of the intermediate or final exam of the course. The practice exam has the same format as the "real" exam, which means that if the practice exam has 20 multiple-choice questions and four free-answer questions, the "real" exam will be the same.